Advances in Geographic Information Science

Series Editors:
Shivanand Balram, Canada
Suzana Dragicevic, Canada

For further volumes:
http://www.springer.com/series/7712

Basudeb Bhatta

Analysis of Urban Growth and Sprawl from Remote Sensing Data

Dr. Basudeb Bhatta
Jadavpur University
Dept. Computer Science & Engineering
Computer Aided Design Centre
Kolkata-700032
India
basu_bhatta@rediffmail.com
basubhatta@gmail.com

ISSN 1867-2434   e-ISSN 1867-2442
ISBN 978-3-642-26287-6   e-ISBN 978-3-642-05299-6
DOI 10.1007/978-3-642-05299-6
Springer Heidelberg Dordrecht London New York

© Springer-Verlag Berlin Heidelberg 2010
Softcover reprint of the hardcover 1st edition 2010
This work is subject to copyright. All rights are reserved, whether the whole or part of the material is concerned, specifically the rights of translation, reprinting, reuse of illustrations, recitation, broadcasting, reproduction on microfilm or in any other way, and storage in data banks. Duplication of this publication or parts thereof is permitted only under the provisions of the German Copyright Law of September 9, 1965, in its current version, and permission for use must always be obtained from Springer. Violations are liable to prosecution under the German Copyright Law.

The use of general descriptive names, registered names, trademarks, etc. in this publication does not imply, even in the absence of a specific statement, that such names are exempt from the relevant protective laws and regulations and therefore free for general use.

Cover design: Bauer, Thomas

Printed on acid-free paper

Springer is part of Springer Science+Business Media (www.springer.com)

Chandrani and *Bagmi*

Preface

Urban growth and sprawl is a pertinent topic for analysis and assessment towards the sustainable development of a city. Environmental impacts of urban growth and extent of urban problems have been growing in complexity and relevance, generating strong imbalances between the city and its hinterland. The need to address this complexity in assessing and monitoring the urban planning and management processes and practices is strongly felt in the recent years.

Determining the rate of urban growth and urban spatial configuration, from remote sensing data, is a prevalent approach in contemporary urban geographic studies. Maps of growth and a classified urban structure derived from remotely sensed data can assist planners to visualise the trajectories of their cities, their underlying systems, functions, and structures. There are currently a number of applications of analytical methods and models available to cities by using the remote sensing data and geographic information system (GIS) techniques, in specific—for mapping, monitoring, measuring, analysing, and modelling.

The international participants are increasingly engaged with the urgent environmental tasks for the sustainable development of their urban regions, the planning challenges faced by the local authorities, and the application of remote sensing data and GIS techniques in the analysis of urban growth to meet these challenges. However, despite the promise of new and fast-developing remote sensing technologies, a gap exists between the research-focused results offered by the urban remote sensing community and the application of these data and methods/models by the governments of urban regions. There is no end of interesting scientific questions to ask about cities and their growth, but sometimes these questions do not match the operational problems and concerns of a given city. This necessitates more focused research and debate in the areas covered by this title.

Kolkata, India   B. Bhatta

Acknowledgments

I am grateful to all the authors of the numerous books and research publications mentioned in the list of references of this book. I have consulted these literatures, picked up the relevant materials, synthesised them and put them in an organised manner with simple language in this book. I express my gratitude to the teachers, researchers and organisations who have contributed a lot to the quality and information content of this book.

I am very much thankful to *Prof Rana Dattagupta*, Director, Computer Aided Design Centre, Computer Science and Engineering Department, Jadavpur University, Kolkata for extending necessary facilities to write this book. I express my gratitude to *Dr Subhajit Saraswati* and *Dr Debashis Bandyopadhyaya*, Construction Engineering Department, Jadavpur University, Kolkata for their help and guidance. I am thankful to my colleagues, especially *Mr Biswajit Giri*, *Mr Chiranjib Karmakar*, *Mr Subrata Das*, *Mr Santanu Glosal*, and *Mr Uday Kumar De*. Without their help and cooperation writing of this book was never possible. I am also thankful to *Mr Joel Newell*, Kingston, Pennsylvania for his contribution of the aerial photograph that is used in the cover of the book.

I would like to express my gratitude to my parents who have been a perennial source of inspiration and hope for me. I also want to thank my wife *Chandrani*, for her understanding and full support, while I worked on this book. My little daughter, *Bagmi*, deserves a pat for bearing with me during this rigorous exercise.

B. Bhatta

Contents

1 Urban Growth and Sprawl 1
   1.1 Introduction 1
   1.2 Urban Geography 1
   1.3 Urban Development, Urban Growth, and Urbanisation 2
   1.4 Urban Area 3
   1.5 Urban Ecosystem 6
   1.6 Urban Sprawl 7
      1.6.1 Defining Urban Sprawl 8
   1.7 Physical Patterns and Forms of Urban Growth and Sprawl ... 10
      1.7.1 Urban Growth Patterns as Sprawl 12
   1.8 Temporal Process of Urban Growth and Sprawl 14

2 Causes and Consequences of Urban Growth and Sprawl 17
   2.1 Introduction 17
   2.2 Causes of Urban Growth and Sprawl 17
      2.2.1 Population Growth 18
      2.2.2 Independence of Decision 20
      2.2.3 Economic Growth 21
      2.2.4 Industrialisation 21
      2.2.5 Speculation 21
      2.2.6 Expectations of Land Appreciation 21
      2.2.7 Land Hunger Attitude 22
      2.2.8 Legal Disputes 22
      2.2.9 Physical Geography 22
      2.2.10 Development and Property Tax 23
      2.2.11 Living and Property Cost 23
      2.2.12 Lack of Affordable Housing 23
      2.2.13 Demand of More Living Space 24
      2.2.14 Public Regulation 24
      2.2.15 Transportation 24
      2.2.16 Road Width 24
      2.2.17 Single-Family Home 25

		2.2.18	Nucleus Family	26
		2.2.19	Credit and Capital Market	26
		2.2.20	Government Developmental Policies	26
		2.2.21	Lack of Proper Planning Policies	26
		2.2.22	Failure to Enforce Planning Policies	26
		2.2.23	Country-Living Desire	27
		2.2.24	Housing Investment	27
		2.2.25	Large Lot Size	27
	2.3	Consequences of Urban Growth and Sprawl		28
		2.3.1	Inflated Infrastructure and Public Service Costs	29
		2.3.2	Energy Inefficiency	30
		2.3.3	Disparity in Wealth	30
		2.3.4	Impacts on Wildlife and Ecosystem	30
		2.3.5	Loss of Farmland	31
		2.3.6	Increase in Temperature	31
		2.3.7	Poor Air Quality	33
		2.3.8	Impacts on Water Quality and Quantity	34
		2.3.9	Impacts on Public and Social Health	34
		2.3.10	Other Impacts	36
3	**Towards Sustainable Development and Smart Growth**			37
	3.1	Introduction		37
	3.2	Sustainable Development		37
	3.3	Smart Growth		39
		3.3.1	Compact Neighbourhoods	42
		3.3.2	Transit-Oriented Development	42
		3.3.3	Pedestrian- and Bicycle-Friendly Design	43
		3.3.4	Others Elements	43
	3.4	Restricting Urban Growth and Sprawl		44
4	**Remote Sensing, GIS, and Urban Analysis**			49
	4.1	Introduction		49
	4.2	Remote Sensing		49
	4.3	Urban Remote Sensing		50
	4.4	Consideration of Resolutions in Urban Applications		52
		4.4.1	Spatial Resolution	53
		4.4.2	Spectral and Radiometric Resolutions	54
		4.4.3	Temporal Resolution	55
	4.5	Geographic Information System		56
		4.5.1	GIS in Urban Analysis	57
	4.6	Urban Analysis		58
		4.6.1	Analysis of Urban Growth	59
		4.6.2	Analysis of Urban Growth Using Remote Sensing Data	61

Contents

5 Mapping and Monitoring Urban Growth 65
  5.1 Introduction 65
  5.2 Image Overlay 65
  5.3 Image Subtraction 68
  5.4 Image Index (Ratioing) 70
  5.5 Spectral-Temporal Classification 72
  5.6 Image Regression 73
  5.7 Principal Components Analysis Transformation 74
  5.8 Change Vector Analysis 75
  5.9 Artificial Neural Network 76
  5.10 Decision Tree 77
  5.11 Intensity-Hue-Saturation Transformation ... 78
  5.12 Econometric Panel 79
  5.13 Image Classification and Post-classification Comparison 79
  5.14 Challenges and Constraints 82

6 Measurement and Analysis of Urban Growth 85
  6.1 Introduction 85
  6.2 Transition Matrices 86
    6.2.1 Transition Matrix in Urban Growth Analysis 86
  6.3 Spatial Metrics 87
    6.3.1 Remote Sensing, Spatial Metrics, and Urban Modelling 88
    6.3.2 Spatial Metrics in Urban Growth Analysis 89
  6.4 Spatial Statistics 92
    6.4.1 Types of Spatial Statistics 92
    6.4.2 Spatial Statistics in Urban Growth Analysis 96
  6.5 Quantification and Characterisation of Sprawl 97

7 Modelling and Simulation of Urban Growth 107
  7.1 Introduction 107
  7.2 Urban Model and Modelling 107
    7.2.1 Classification of Urban Models 109
  7.3 Theoretical Models 109
  7.4 Aggregate-Level Urban Dynamics Models 110
  7.5 Complexity Science-Based Models 110
    7.5.1 Cell-Based Dynamics Models 111
    7.5.2 Agent-Based Models 113
    7.5.3 Artificial Neural Network (ANN)-Based Models 114
    7.5.4 Fractal Geometry-Based Models 116
  7.6 Rule-Based Land-Use and Transport Models ... 116
  7.7 Modelling of Urban Growth 118
    7.7.1 Modelling of Urban Growth from Remote Sensing Data 118

8 Limitations of Urban Growth Analysis 123
  8.1 Introduction 123

	8.2	Data and Scale Dependency	123
		8.2.1 Spatial Scale	124
		8.2.2 Pattern Quantification Scale	125
		8.2.3 Pattern Summarisation Scale	126
	8.3	Data Generation Methods	126
	8.4	Classification Accuracy	126
	8.5	Selection of Metrics	127
	8.6	Definition of the Spatial Domain	128
	8.7	Spatial Characterisation	130
	8.8	Spatial Dependency (Autocorrelation)	131
	8.9	Spatial and Temporal Sampling	132
	8.10	Modifiable Areal Unit Problem	132

References . . . 135

Index . . . 169

About the Book

This book focuses on available methods and models for the analysis of urban growth and sprawl from remote sensing data along with their merits and demerits. However, this book does not aim to explain the concepts of these methods and models elaborately; rather it attempts to mention applications of these methods and models along with brief descriptions to gain a perception of their conceptual differences and how they are applicable in urban growth analysis. Therefore, the comparison among these methods and models does not result in absolute acceptance or rejection of any of these methods or models; rather it aims to answer this important question that to what extent does any of these methods or models fit in the analysis of urban growth and sprawl. The strategy that this book follows is reviewing studies in spatio-temporal analysis of urban growth and sprawl, and identifying the merits and demerits of each method or model. Although the focus is on analysis, as it should be; urban growth and sprawl—their patterns, process, causes, consequences, and countermeasures are also clearly described in context.

This book begins with discussion on urban growth and sprawl. As readers go through the chapters of this book, they will learn about the applications of remote sensing and GIS in the discipline of urban growth and sprawl—mapping, measuring, analysing, and modelling. The primary purpose of this book is to be a rigorous literature review for the scientists and researchers. This book will also be appreciated by the academicians for preparing lecture notes and delivering lectures. Post graduate students of urban geography or urban/regional planning may refer this book as additional studies. Industry professionals may also be benefited from the discussed methods and models along with numerous citations. It is hoped that this book will also attract and inspire individuals who might consider a specialised career in the field of urban planning and sustainable development or in the broader fields allied with urban growth.

Content and Coverage

The book is comprised with eight chapters; each chapter commences with an introduction that briefly outlines the topics covered in the respective chapter. The chapters also contain illustrations, notes and numerous citations that complement the text.

Chapter 1 covers the basic concepts of urban growth and sprawl along with some allied issues, especially, details on their patterns and process. Chapter 2 describes the causes and consequences of urban growth and sprawl in brief. Chapter 3 introduces the sustainable development and smart growth, and available approaches to restrict urban growth and sprawl towards achieving the sustainability.

Chapter 4 deals with general discussion on remote sensing data and GIS techniques for urban analysis. Chapter 5 describes available methods for mapping and monitoring of urban growth along with respective merits and demerits. Different measurement and analytical techniques for urban growth and sprawl are furnished in Chap. 6. Chapter 7 houses modelling and simulation techniques of urban growth; and Chap. 8 is aimed to discuss several limitations of urban growth analysis from remote sensing data.

A very rich list of references at the end, which have been referred in the text suitably, is for the benefit of the readers for further readings.

Acronyms

ANN	Artificial Neural Network
CA	Cellular Automaton (Cellular Automata)
CAST	City Analysis Simulation Tool
CUF	California Urban Futures
CVA	Change Vector Analysis
DUEM	Dynamic Urban Evolutionary Model
ETM	Enhanced Thematic Mapper
FCAUGM	Fuzzy Cellular Automata Urban Growth Model
GDP	Gross Domestic Product
GIS	Geographic Information System
GNSS	Global Navigation Satellite System
GWR	Geographically Weighted Regression
HIS	Intensity-Hue-Saturation
IFOV	Instantaneous Field of View
IRS	Indian Remote Sensing
LEAM	Land-use Evolution and Impact Assessment Modelling
LISS	Linear Imaging Self-scanning Sensor
MAD	Minimum Average Distance
MAUP	Modifiable Areal Unit Problem
MSS	MultiSpectral Scanner
NASA	National Aeronautics and Space Administration
NDVI	Normalized Differenced Vegetation Index
NIR	Near-InfraRed
NN	Neural Network
NOAA	National Oceanic and Atmospheric Administration
OPUS	Open Platform for Urban Simulation
PC	Principal Component
PCA	Principal Components Analysis
PDR	Purchase of Development Rights
RGB	Red-Green-Blue
SLEUTH	Slope, Land-use, Exclusion, Urban extent, Transportation and Hillshade
SMSA	Standard Metropolitan Statistical Area

SNN	Simulated Neural Network
SPOT	Satellite Pour l'Observation de le Terre
SVM	Support Vector Machines
TM	Thematic Mapper
UES	Urban Expansion Scenario
UGB	Urban Growth Boundary
US	United States
USA	United States of America

Chapter 1
Urban Growth and Sprawl

1.1 Introduction

Urban geography is the study of urban areas in terms of population concentration, infrastructure, economy and environmental impacts. The study of urban geography has evinced interest from a wide range of experts. The multidisciplinary gamut of the subject invokes the interest from ecologists, to urban planners and civil engineers, to sociologists, to administrators and policy makers, and the common people as well. This is because of the multitude of activities and processes that take place in the urban ecosystems everyday.

Study of urban growth is a branch of urban geography that concentrates on cities and towns in terms of their physical and demographic expansion. Urban sprawl, an undesirable type of urban growth, is one of the major concerns to the city planners and administrators. In the recent decades, analysis of urban growth from various perspectives has become an essentially performed operation for many reasons.

Before we proceed to concentrate on analytical techniques of urban growth and sprawl, we should have an overall concept on urban growth and sprawl. This chapter is aimed to discuss urban growth and sprawl in addition to some allied issues. Specifically, this chapter furnishes the patterns and processes of urban growth and sprawl.

1.2 Urban Geography

Geography is the scientific study of [geo]spatial pattern and process. It seeks to identify and account for the location and distribution of human and physical phenomena on the earth's surface. Emphasis in geography is placed upon the organisation and arrangement of phenomena, and upon the extent to which they vary from place to place and time to time. Although it shares a subtractive interest in the same phenomena as other social and environmental sciences, the spatial perspective upon phenomena which is adopted in geography is distinctive. No other discipline has location and distribution as its major focus of study. It is the characteristics of space as a dimension, rather than the properties of phenomena which are located in space,

B. Bhatta, *Analysis of Urban Growth and Sprawl from Remote Sensing Data*,
Advances in Geographic Information Science, DOI 10.1007/978-3-642-05299-6_1,
© Springer-Verlag Berlin Heidelberg 2010

that is of central and overriding concern. Geography aims to develop general rather than unique explanations. It proceeds from the assumption that there is basic regularity and uniformity in the location and occurrence of phenomena and that this order can be identified and accounted for by geographical analysis. In examining spatial structure, geography focuses upon those distributional characteristics that are common to a wide range of phenomena. To this end, emphasis in geographical study is placed upon *models and theories of locations* rather than upon descriptions of individual features (Clark 1982). The overriding aim is to develop an understanding of general principles which determine the location of human and physical characteristics.

Urban geography is a branch of geography that concentrates upon the location and spatial arrangement of towns and cities and their evolutions. In simple words, urban geography is the study of urban areas. It can be considered a part of the larger field of *human geography*. Human geography is a branch of geography that focuses on the study of patterns and processes that shape human interaction with the built environment, with particular reference to the causes and consequences of the spatial distribution of human activity on the earth's surface. However, urban geography can often overlap with other fields such as anthropology and urban sociology. It seeks to add a spatial dimension to our understanding of urban places and urban problems. Urban geographers are concentrated to identify and account for the distribution evolution of towns and cities, and the spatial similarities or contrasts that exit within and between them. They are concerned with both the contemporary urban pattern and with the ways in which the distribution and internal arrangement of towns and cities have changed over time. Urban geographers also look at the development of settlements. Therefore, it involves planning of city expansion and improvements. Emphasis in urban geography is directed towards the understanding of those social, economic and environmental processes that determine the location, spatial arrangement, and evolution of urban places. In this way, urban geographical analysis supplements and complements the insights provided by allied disciplines in the social and environmental sciences which recognise the city as a distinctive focus of study.

1.3 Urban Development, Urban Growth, and Urbanisation

Urban development is the process of emergence of the world dominated by cities and by urban values (Clark 1982). The occurrence of urban development is so general, and its implications are so wide, that it is possible to view much of recent social and economic history in terms of the attempts to cope with its varying consequences. The rise of great cities and their growing spatial influence initiated a change from largely rural to predominantly urban places and patterns of living that has affected most countries over the last two centuries. Currently, not only do large numbers of people live in or immediately adjacent to towns and cities, but whole segments of the population are completely dominated by urban values, expectations and life styles. From its origins as a locus of non-agricultural employment, the city has become the major social, cultural and intellectual stimulus in modern urban society.

It is important to draw a clear distinction between the two main processes of urban development—*urban growth* and *urbanisation*. According to Clark (1982), *urban growth* is a spatial and demographic process and refers to the increased importance of towns and cities as a concentration of population within a particular economy and society. It occurs when the population distribution changes from being largely hamlet and village based to being predominantly town and city dwelling. *Urbanisation*, on the other hand, is aspatial (non-spatial) and social process which refers to the changes of behaviour and social relationships that occur in social dimensions as a result of people living in towns and cities. Essentially, it refers to the complex change of life styles which follow from the impact of cities on society.

However, nowadays, the word 'urbanisation' is commonly used in more broad sense and it refers to the physical growth of urban areas from rural areas as a result of population immigration to an existing urban area. Effects of urbanisation include change in urban density and administration services. Urbanisation is also defined as 'movement of people from rural to urban areas with population growth equating to urban migration' (United Nations 2005a). As more and more people leave villages and farms live in cities, urban growth results. Important to realise that, the urbanisation process refers to much more than simple population growth; it involves changes in the economic, social and political structures of a region. Rapid urbanisation is responsible for many environmental and social changes in the urban environment and its effects are strongly related to global change issues. The rapid growth of cities strains their capacity to provide services such as energy, education, health care, transportation, sanitation and physical security. However, direct implication of urbanisation is attributed to spatial growth of towns and cities or in other words growth in urban areas that is commonly referred to as 'urban growth'.

The spatial configuration and the dynamics of urban growth are important topics of analysis in the contemporary urban studies. Several studies have addressed these issues with or without the consideration of demographic process and urbanisation which have dealt with diverse range of themes (e.g., Acioly and Davidson 1996; Wang et al. 2003; Páez and Scott 2004; Zhu et al. 2006; Belkina 2007; Puliafito 2007; Yanos 2007; Martinuzzi et al. 2007; Hedblom and Soderstrom 2008; Zhang and Atkinson 2008; Geymen and Baz 2008).

1.4 Urban Area

Urban area commonly refers to *towns* and *cities*—an *urban landscape*. The definition of urban area changes from country to country. There are various ways to define what is urban and what is part of an urban area (Carter 1981). In Britain, open space that is completely surrounded by the other urban land-use types (e.g. residential, industrial, and commercial, etc.) belongs to an urban area (Carter 1981). In China, collective-owned nursery land may be defined as farmland even when it is completely surrounded by urban land-use types (Li 1991). Similarly, it is often a subjective matter to decide whether a lake or coastal waters within or beside an urban area should be allocated to the urban or not (Huang et al. 2007).

This confusion can be eliminated if the *urban land-use* is differentiated from *urban land-cover*. 'Land-cover' corresponds to the physical condition of the ground surface, for example, forest, grassland, cropland, and water; while 'land-use' reflects human activities such as the use of the land, for example, industrial, residential, recreational, and agricultural. Land-cover refers to features of land surface, which may be natural, semi-natural, managed, or manmade. They are directly observable by a remote sensor. Whereas, land-use refers to activities on land, or classification of land according to how it is being used. Not always directly observable by a remote sensor, inferences about land-use can often be made from land-cover. Practices are there to consider urban (or urban area) both as land-use and land-cover. 'Urban' pixels that form the basis of many remote sensing analyses consist typically of developed and impervious[1] surfaces (pixels) that include built structures, concrete, asphalt, runways, and buildings. In this case it is land-cover that is often labelled as *built-up area* or *built-up cover* or *urban cover*. However, from the perspective of land-use, urban areas 'may also include a number of non-developed and vegetated pixels such as parklands and urban forests, and may exclude developments that are components of other land-uses' (Martinuzzi et al. 2007). In this sense urban area refers to urban landscape—towns and cities.

Although determining the urban area as land-cover is a straightforward approach (whether it is impervious or not), however, determining the urban area as land-use is a difficult task. An urban to rural transect shows a gradient and makes the process of delineating a boundary between rural and urban more complex. Cities (or towns) have administrative boundaries associated with them in the sense that city governments have jurisdiction over certain well-defined administrative areas. But the area contained within the jurisdictional boundary of a city has little to do with the metropolitan area of the city. In some cases, this area is very small in comparison with the size of the metropolitan area. The Los Angeles metropolitan area, for example, contains 35 independent municipalities; the Kolkata metropolitan area contains 38 independent municipalities and three municipal corporations. In other cases, say in Beijing, the jurisdictional boundary of the municipality contains an area that is much larger than the built-up area of the city. The official area of the municipality is therefore not a very precise measure, neither of the built-up area of the city (urban land-cover) nor of what we intuitively grasp to be the city (urban land-use) (Angel et al. 2005). Furthermore, the extent of a city is a dynamic phenomenon; it changes over time. The jurisdictional boundary of the city can not be changed frequently owing to administrative complexities.

---

[1] Impervious surfaces are mainly constructed surfaces—rooftops, sidewalks, roads, runways, and parking lots—covered by impenetrable materials such as asphalt, concrete, and stone. These materials effectively seal surfaces, repel water and prevent precipitation and meltwater from infiltrating soils.

1.4 Urban Area

There are several approaches to determine the boundary (or extent) of a city or town. Traditionally the physical delimitation of urban areas and agglomerations has been characterised by two clearly differentiated approaches. On the one hand delimitation is based upon physical or morphological criteria, where the continuous built-up area, or the density of contiguous ambits, comprises the basic mechanisms for the delimitation. On the other hand, studies based upon functional or economic criteria, where the emphasis is placed upon the existing relations and flows throughout the urbanised territory where the relation between place of residence and place of work is fundamental (Roca et al. 2004). Among these two, the former is more practiced. Population density is also another preferred option of defining urban area. According to U.S. Census Bureau (2000) all territory, population, and housing units located within census block that have a population density of at least 390 people/km^2, plus surrounding census block that have an overall density of at least 195 people/km^2, are considered as urban areas. All territory, population, and housing units located outside of urban areas are rural.

Delineation of morphological urban area includes some of the earliest work using Landsat sensors (Forster 1983, 1985). This has since been followed up with more sophisticated approaches to settlement identification. These range from simple indices (Gao and Skillcorn 1998; Zha et al. 2003), to regional characterisations (Vogelmann et al. 1998; Civco et al. 2002), and to complex data fusion approaches using *Bayes modeling* and *Markov random fields* (Yu et al. 1999). Techniques of urban mapping have further progressed to extracting statistical measures of characterisation. These include deriving simple measures of urban morphology from satellite sensor imagery (Mesev et al. 1995; Webster 1996; Longley and Mesev 2001) to the use of texture information (Pesaresi and Bianchin 2001) and geostatistics (Brivio and Zilioli 2001) or spatial statistics (Ramos and Silva 2003) for urban structure and pattern characterisation and thereby delineation of urban area.

In addition to the popular optical multispectral sensors, alternative satellite sensors are also being used increasingly in morphological urban area delineation. The additional discriminating power over optical sensors of imaging radar has led to its use in imaging urban areas (Taket et al. 1991; Hepner et al. 1998). A more novel approach to mapping urban extent is that of satellite radar interferometry, by which the degree of coherence between a pair of radar images is measured (Grey and Luckman 2003). Nighttime imagery has also found its place in this field. Nighttime imagery that shows city lights can be used to estimate settlement area (Imhoff et al. 1997). Bhatta (2009b) has shown an interesting method to determine the natural extent of urban area from remote sensing data, which suggests conducting a neighbourhood search to derive the contiguous urban area from a binary classified image. Although the general idea of what is 'urban' varies from census and remote sensing environment, they are not mutually exclusive.

There are also difficulties in characterising the urban area—whether it is a town, or city, or megacity, or just rural developed land. In general, there are no standards, and each country develops its own set of criteria for distinguishing cities and towns. A town is a type of settlement ranging from a few to several thousand inhabitants; the precise meaning varies between countries and is not always a matter of legal definition. Usually, a 'town' is thought of as larger than a village but smaller than a 'city'. A city is generally defined as a political unit, i.e., a place organised and governed by an administrative body. A way of defining a town/city or an urban area is by the number of residents or settlements. The United Nations defines areas having settlements of over 20,000 as urban, and those with more than 100,000 as cities (United Nations 1994). Therefore, 'town' may be defined as an area of having settlements between 20,000–100,000. The United States defines an urbanised area as a city and surrounding area, with a minimum population of 50,000. A metropolitan area includes both urban areas and rural areas that are socially and economically integrated with a particular city. Generally, cities with over 5 million inhabitants are known as megacities.

1.5 Urban Ecosystem

An urban landscape may be viewed as a *system* (often called *urban ecosystem*) that integrates physical, social, economic, ecological, environmental, infrastructure and institutional subsystems; where urban growth and sprawl is an outcome of change in performance/functioning of these subsystems.

System is an assemblage of entities/objects, real or abstract, comprising a whole with each and every component/element interacting or related to another one. Any element which has no relationship with any of the elements of a system is not a component of that system. A system may be defined with simple words as 'a group of connected entities and activities which interact for a common purpose'. For example, a car is a system in which all the components operate together to provide transportation. Some other examples are natural systems (like the ecosystem, blood system, solar system, etc.), manmade system (machines, industrial plants, telecommunication infrastructure networks, computer storage systems, etc.), abstract systems (conceptual modelled systems like traffic system models, computer programs, etc.), and descriptive and normative systems (man and other living system activity, plans, bus/train timetable, ethical systems, etc.). Every division or aggregation of real objects/entities into systems is arbitrary; therefore, each of the systems is a subjective abstract concept (Bhatta 2008). Urban ecosystem is a hybrid system that perhaps combines natural, manmade, descriptive and normative systems.

Urban ecosystems are the consequences of the intrinsic nature of humans as social beings to live together in towns and cities. Thus when the early humans evolved they settled on the banks of the rivers that dawned the advent of civilisations. An inadvertent increase in the population complimented with creativity, humans were able to invent wheel and light fire, created settlements and started lived in forests too. Gradually, with the development of their communication skills by the form of languages through speech and script, the humans effectively utilised this to make enormous progress in their life styles. All this eventually led to the initial human settlements into villages, towns and then into cities. In the process humans now live in complex ecosystems called urban ecosystems.

Important to realise, an unprecedented population growth and migration, an increased urban population, and urbanisation are inadvertent. More and more towns and cities bloomed with a change in the land-use and land-cover along the myriad of landscapes and ecosystems found on earth. Today, humans can boast of living under a wide range of climatic and environmental conditions. This has further led to humans contributing the urban centres at almost every corner of the earth. These urban ecosystems are a consequence of urban development through rapid industrial centres and blooming up of residential colonies, also became hub of economic, social, cultural, and political activities.

1.6 Urban Sprawl

In the late 1950s, urbanised areas in USA have extended outside rapidly during the suburbanisation process of residence, industry and commerce, which encroached large amount of farmland and forest, brought negative effects to environment and caused more traffic problems. This pattern of urban development out of control has been regarded as *urban sprawl* (Zhang 2004). Although sprawl has existed to some extent since the invention of the automobile, but in truth it rose up during the post-World War II housing boom (Duany et al. 2001). Later, in general, most of all cities have experienced or are experiencing the sprawl; including the cities in developing countries.

The concept of sprawl-emergence of a situation of unauthorised and unplanned development, normally at the fringe areas of cities especially *haphazard* and *piecemeal construction* of homesteads, commercial areas, industrial areas and other non-conforming land-uses, generally along the major lines of communications or roads adjacent to specified city limits, is observed which is often termed as the urban sprawl. The area of urban sprawl is characterised by a situation where urban development adversely interferes with urban environment which is neither an acceptable urban situation nor suitable for an agricultural rural environment (Rahman et al. 2008).

Urban sprawl has aroused wide social focus because it can impede regional sustainable development. An empirical research, conducted by Bengston et al. (2005), shows that public concern about the impacts of sprawling development patterns has grown rapidly during the latter half of the 1990s. Related studies have come out consequently which mainly cover *patterns, process, causes, consequences*, and *countermeasures*.

1.6.1 Defining Urban Sprawl

Sprawl [2] as a concept suffers from difficulties in definition (Johnson 2001b; Barnes et al. 2001; Wilson et al. 2003; Roca et al. 2004; Sudhira and Ramachandra 2007; Angel et al. 2007). Galster et al. (2001) critiqued the conceptual ambiguity of sprawl observing that much of the existing literature is 'lost in a semantic wilderness'. Their review of the literature found that sprawl can alternatively or simultaneously refer to: (1) certain patterns of land-use, (2) processes of land development, (3) causes of particular land-use behaviours, and (4) consequences of land-use behaviours. They have reviewed many definitions of sprawl from different perspectives. It seems that sprawl is used both as a *noun* (condition) and *verb* (process); and suffers from a lack of clarity even though many would claim to 'know it when they see it' (Galster et al. 2001).

Urban sprawl is often discussed without any associated definition at all. Some authors make no attempt at definition while others 'engage in little more than emotional rhetoric' (Harvey and Clark 1965). Johnson (2001b) presents several alternative definitions for consideration, concluding that there is no common consensus. Because sprawl is demonised by some and discounted by others, how sprawl is defined depends on the perspective of who presents the definition (Barnes et al. 2001). Wilson et al. (2003) argues that the sprawl phenomenon seeks to describe rather than define. Galster et al. (2001) also emphasised describing the sprawl rather than defining. However, let us review some of the definitions of urban sprawl in the existing literature to understand its complexity.

One of the ways the Oxford Dictionary (2000) defines the sprawl as

> A large area covered with buildings, which spreads from the city into the countryside in an ugly way.

Ottensmann (1977) has identified urban sprawl as

> The scattering of new development on isolated tracts, separated from other areas by vacant land.

It is also often referred to as *leapfrog development* (Gordon and Richardson 1997a).

The Sierra Club (2001) describes suburban sprawl as

> ...irresponsible, often poorly-planned development that destroys green space, increases traffic and air pollution, crowds schools and drives up taxes.

Others compare sprawl to the disease process, calling it *a cancerous growth* or *a virus* (DiLorenzo 2000). Less strident descriptions include 'the scattering of urban settlement over the rural landscape' (Harvey and Clark 1965), 'low-density urbanisation' (Pendall 1999), 'the land-consumptive pattern of suburban development' (Wilson et al. 2003) and 'discontinuous development' (Weitz and Moore 1998).

---

[2] Although the literal meaning of sprawl varies across, in this literature, *urban sprawl* and *sprawl* have been used interchangeably.

However, sprawl must be considered in a space-time context as not simply the increase of urban lands in a given area, but the rate of increase relative to population growth (Barnes et al. 2001). United States Environmental Protection Agency defined sprawl in 2001 as

> At a metropolitan scale, sprawl may be said to occur when the rate at which land is converted to non-agricultural or non-natural uses exceeds the rate of population growth (cited in Barnes et al. 2001).

Sudhira and Ramachandra (2007) have stated

> Urban sprawl refers to the outgrowth of urban areas caused by uncontrolled, uncoordinated and unplanned growth. This outgrowth seen along the periphery of cities, along highways, and along roads connecting a city, lacks basic amenities like sanitation, treated water supply, primary health centre, etc. as planners were unable to visualise such growth during planning, policy and decision-making.

Although accurate definition of urban sprawl is debated, a general consensus is that urban sprawl is characterised by unplanned and uneven pattern of growth, driven by multitude of processes and leading to inefficient resource utilization. The direct implication of such sprawl is change in land-use and land-cover of the region as sprawl induces the increase in built-up and paved area (Sudhira and Ramachandra 2007).

It is worth mentioning that opinions on sprawl held by researchers, policy makers, activists, and the public differ sharply; and the lack of agreement over how to define sprawl undoubtedly complicates efforts to restrict this type of land development. This necessitates a clear definition of urban sprawl without characterising its causes or attributes, and can be given as:

> Urban sprawl is the less compact outgrowth of a core urban area exceeding the population growth rate and having a refusal character or impact on sustainability of environment and human.

To describe the preceding definition, non-sprawling development means: (1) discouraging outgrowth; (2) in case of any unavoidable outgrowth, it should be high compact to reduce the city-size; (3) urban growth-rate should not exceed the population growth rate of that area; and (4) should not harm the interest and need of human and environment much, in the context of present and future. Fulfilment of these expectations tends to a smart and sustainable urban growth, failing of which results in sprawl.

The given definition is short and simple. But unfortunately it lacks clear insight, for example, the quantity and quality of negative impact, or the magnitude of compactness in connection of 'less compact'. This certainly gives some short of freedom to stretch this definition; however, in some of the cases this freedom may also be desired. For example, neither all cities are equally compact nor they can be compacted equally. Therefore the degree of compactness can be considered according to the application area, and debates by the proponents can help in determining the degree of compactness for a specific area.

Varying characterisation of sprawl (after Torrens 2008)	Growth	Social	Aesthetic	Decentralisation	Accessibility	Density	Open space	Dynamics	Costs	Benefits
Audirac et al. (1990)		•								
Bae and Richardson (1994)						•				•
Benfield et al. (1999)									•	
Burchell et al. (1998)	•		•		•	•			•	
Calthorpe et al. (2001)			•							
Clapham Jr. (2003)								•		
Duany et al. (2001)			•							
El Nasser and Overburg (2001)						•				
Ewing (1997)		•		•	•	•		•	•	•
Ewing et al. (2002)		•		•	•	•	•	•	•	•
Farley and Frey (1994)		•								
Galster (1991)		•		•						
Galster et al. (2001)	•			•						
Gordon and Richardson (1997a)						•		•		•
Gordon and Richardson (1997b)						•		•		•
Hasse and Lathrop (2003a)							•			
Hasse and Lathrop (2003b)			•	•	•					
Hasse (2004)							•			
HUD (1999)							•			
Lang (2003)						•				
Ledermann (1967)						•				
Lessinger (1962)								•		
Malpezzi (1999)						•				
OTA (1995)						•				
Peiser (1989)						•				
Pendall (1999)						•				
RERC (1973)						•			•	
Sierra Club (1998)				•						
Sudhira et al. (2004)								•		

1.7 Physical Patterns and Forms of Urban Growth and Sprawl

Urban growth, *urban expansion* and *urban sprawl* are sometimes used synonymously by the common people, although they are different. Urban *growth* is a sum of increase in developed land. One of its forms is *expansion*. Whereas, urban growth having some special characteristics (typically has a negative connotation) is *sprawl*.

Urban growth can be defined in relation to Forman's (1995) landscape transformation processes. Although the processes themselves are quite similar, the urban growth defines growth from the perspective of a growing urban patch while the landscape transformation processes define fragmentation types as a reduction of non-developed land-cover types. Pattern of urban land-use/cover refers to the arrangement or spatial distribution of built environment.

Wilson et al. (2003) have identified three categories of urban growth: *infill*, *expansion*, and *outlying*, with outlying urban growth further separated into *isolated*, *linear branch*, and *clustered branch* growth (Fig. 1.1). The relation (or distance) to

1.7 Physical Patterns and Forms of Urban Growth and Sprawl

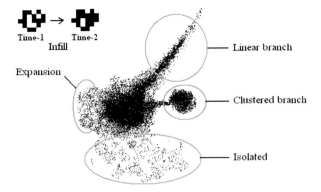

Fig. 1.1 Schematic diagram of urban growth pattern

existing developed areas is important when determining what kind of urban growth has occurred.

An *infill* growth is characterised by a non-developed pixel being converted to urban use and surrounded by at least 40% existing developed pixels. It can be defined as the development of a small tract of land mostly surrounded by urban land-cover (Wilson et al. 2003). Ellman (1997) defines infill policies as the encouragement to develop vacant land in already built-up areas. Infill development usually occurs where public facilities such as sewer, water, and roads already exist (Wilson et al. 2003). Forman (1995) describes *infill attrition* as the disappearance of objects such as patches and corridors.

An *expansion* growth is characterised by a non-developed pixel being converted to developed and surrounded by no more than 40% existing developed pixels. This conversion represents an expansion of the existing urban patch (Wilson et al. 2003). Expansion-type development has been called *metropolitan fringe development* or *urban fringe development* (Heimlich and Anderson 2001; Wasserman 2000). Forman (1995) discusses it as *edge development*, defined as a land type spreading unidirectionally in more or less parallel strips from an edge. The analogous land transformation is *shrinkage*, defined as the decrease in size of objects, such as patches (Forman 1995).

Outlying growth is characterised by a change from non-developed to developed land-cover occurring beyond existing developed areas (Wilson et al. 2003). This type of growth has been called development beyond the urban fringe (Heimlich and Anderson 2001). The outlying growth designation is broken down into the following three classes: *isolated*, *linear branch*, and *clustered branch* (Wilson et al. 2003).

Isolated growth is characterised by one or several non-developed pixels some distance from an existing developed area being developed. This class of growth is characteristic of a new house or similar construction surrounded by little or no developed land (Wilson et al. 2003). Forman (1995) defines it as *perforation*, which is the process of making holes in an object such as a habitat.

Linear branch can be defined as an urban growth such as a new road, corridor, or a new linear development that is generally surrounded by non-developed land

and is some distance from existing developed land. A linear branch is different from isolated growth in that the pixels that changed to urban are connected in a linear fashion (Wilson et al. 2003). Forman (1995) defines it as *corridor*, means a new corridor such as a road that bisects the initial land type. Forman (1995) defines two land transformation processes that apply to linear branch. The first is *dissection*, defined as the carving up or subdividing of an area using equal-width lines. The second is *fragmentation*, defined as the breaking up of a habitat or land type into smaller parcels. Fragmentation also applies to clustered branch. *Clustered branch* defines a new urban growth that is neither linear nor isolated, but instead, a cluster or a group. It is typical of a large, compact, and dense development (Wilson et al. 2003).

1.7.1 Urban Growth Patterns as Sprawl

Wilson et al. (2003) have not attempted to characterise the sprawl, arguing that creating an urban growth model instead of an urban sprawl model allows us to quantify the amount of land that has changed to urban uses, and lets the user decide what he or she considers to be urban sprawl. This argument, no doubt, discourages towards understanding the sprawl and efforts to make it standardise to model one or more scale metrics to measure the sprawl. However, standardisation of sprawl is not an easy task. It is important to realise that every type of urban growth are not considered as sprawl; and the one development which can be considered as urban sprawl by someone may not be considered as sprawl by others (Roca et al. 2004). Furthermore, urban sprawl has a negative connotation on the society and environment, and not all urban growth is necessarily unhealthy. In fact, some types of urban growth (e.g., 'infill' growth) are generally considered as remedies to urban sprawl. Therefore, sprawl can not be characterised by the simple quantification of the amount of land that has changed to urban uses. Sprawl phenomenon should be treated separately than the general urban growth.

Harvey and Clark (1965) have identified three major forms of urban sprawl. First, *low-density continuous development sprawl*, second is *ribbon development sprawl*, and third is *leap-frog development sprawl*. These are basically expansion growth, linear branch growth, and clustered (as well as isolated) branch growth accordingly as defined by Wilson et al. (2003). Angel et al. (2007) also identified three basic forms of urban growth as sprawl—(a) *a secondary urban centre*, (b) *ribbon development*, and (c) *scattered development*.

Developmental patterns frequently characterised as sprawl also include *low density* (Lockwood 1999), *random* (USGAO 1999), *large-lot single-family residential* (Popenoe 1979), *radial discontinuity* (Mills 1980), *single land-use* or *physical separation of land-uses* (Cervero 1991), *widespread commercial development* (Downs 1999), *strip commercial* (Black 1996), *peripheral urban development with progressively increased land consumption* (Roca et al. 2004), and *non-compact* (Gordon and Richardson 1997a) amongst others.

1.7 Physical Patterns and Forms of Urban Growth and Sprawl

An important type of urban growth—*infill*, generally not considered as sprawl. Most of the proponents encourage infill to encounter the sprawl. However, infill should not be considered as simple as it defines. Rather it should be judged and justified with multi-criteria circumstances, for example the growth rate of population in consideration or the actual need of infill by sacrificing what. Furthermore, 100% built-up saturation or urban congestion will also have negative impacts on urban ecosystem.

Ewing (1994) reviewed several patterns of urban sprawl; and argued that the pattern of sprawl is 'like obscenity', the experts may know sprawl when they see it. He (she) has identified two problems with the archetypes. First, sprawl is a 'matter of degree'. The line between scattered development and so-called *polycentric* [3] (*multinucleated*) development is a fine one. 'At what number of centres polycentrism ceases and sprawl begins is not clear' (Gordon and Wong 1985). Scattered development is classic sprawl; it is inefficient from the standpoints of infrastructure and public service provision, personal travel requirements, and the like. Polycentric development, on the other hand, is more efficient than even compact and centralised (*monocentric*) development when metropolitan areas grow beyond a certain size threshold (Haines 1986). A polycentric development pattern permits clustering of land-uses to reduce trip-lengths without producing the degree of congestion extant in a compact centralised pattern (Gordon et al. 1989).

In a similar way, the line between leapfrog development and economically efficient 'discontinuous development' is not always clear. Leapfrog development may occur naturally due to variations in terrain; some portions of the land may not be suitable for urban development (Harvey and Clark 1965). New communities nearly always start up just beyond the urban fringe, where large tracts of land are available at moderate cost (Ewing 1991). Some sites are necessarily bypassed in the course of development, awaiting commercial or higher-density residential uses that will become viable after the surrounding area matures (Ohls and Pines 1975). An important question on sprawl may be, 'how long is required for compaction?' as opposed to whether or not compaction occurs at all (Harvey and Clark 1965). Whether leapfrog development is inefficient will depend on how much land is bypassed, how long it is withheld, how it is ultimately used, and the nature of leapfrog development (Breslaw 1990).

Compactness also does not have a generally accepted definition (Tsai 2005). Gordon and Richardson (1997a) defined compactness as high-density or monocentric development. Ewing's definition (1997) was some concentration of employment

---

[3] Polycentrism is the principle of organisation of a region around several political, social or financial centres. A region is said to be polycentric if it is distributed almost evenly among several centres in different parts of the region. Monocentrism, in contrast, is the organisation of a region's political, social or financial activities centred in a single part (generally the central business district) of the region (Arnott and McMillen 2006). Modern cities are mainly polycentric. Polycentric cities are the outcome of the urbanisation and they have several advantages, for example, they reduce the traffic congestion at the city-centre, reduce the travel time as because there are multiple city-centres, and distribute the financial activities.

and housing, as well as some mixture of land-uses. Alternatively, Anderson et al. (1996) defined both monocentric and polycentric forms as being compact. Other definitions are more measurement-based. Bertaud and Malpezzi (1999) developed a compactness index—the ratio between the average distance from home to central business district, and its counterpart in a hypothesised cylindrical city with equal distribution of development. Galster et al. (2001) described compactness as the degree to which development is clustered and minimises the amount of land developed in each square unit of area. This clearly indicates that there is no consensus regarding the compactness. One common theme is the vague concept that compactness involves the concentration of development; however, the degree of compactness is not well understood.

The difference between strip development and other linear patterns is also a matter of degree. Likewise, the difference between low-density urban development, exurban development and rural residential development is also a matter of degree. Where to draw the line between sprawl and related forms of efficient development will be subject to challenge unless the analysis is based on impacts. It is the impacts of development that present development patterns as undesirable, not the patterns themselves (Ewing 1994).

The second problem with the archetypes is that sprawl has multiple dimensions, which are glossed over in the simple constructs. It is sometimes said that growth management has three dimensions—*density*, *land-use*, and *time*. The same is true of sprawl. Leapfrog development is a problem only in the time dimension; in terms of ultimate density and land-use, leapfrog development may be relatively efficient (Ewing 1994).

Similarly, low-density development is problematic in the density dimension, and strip development in the land-use dimension (since it consists mainly of commercial uses (Black 1996)). If development is clustered at low gross densities, or land-uses are mixed in a transportation corridor, these patterns become relatively efficient (Ewing 1994). Again, it is the impacts of development that render development patterns desirable or undesirable, not the patterns themselves.

Studies analysing the costs of alternative development patterns have, by necessity, defined alternatives in multiple dimensions, rather than limiting themselves to simple sprawl archetypes. One cannot analyse the costs of leapfrog development, for example, without specifying densities and land-uses. Ultimately, what distinguishes sprawl from alternative development patterns is the degree of these dimensions. Numerous approaches have been demonstrated by the researchers to quantify the degree of these dimensions and to characterise sprawl which will be discussed in Chap. 6.

1.8 Temporal Process of Urban Growth and Sprawl

Urban sprawl should be considered both as a *pattern* of urban land-use—that is, a spatial configuration of a metropolitan area in a temporal instant—and as a *process*, namely as the change in the spatial structure of cities over time. Sprawl as a

1.8 Temporal Process of Urban Growth and Sprawl

pattern or a process is to be distinguished from the causes that bring such a pattern about, or from the consequences of such patterns (Galster et al. 2001). If the sprawl is considered as a pattern it is a static phenomenon and as a process it is a dynamic phenomenon. Some of the researchers have considered sprawl as a static phenomenon, whereas some have analysed it as a dynamic phenomenon; however, most of the researchers shout for both.

Sprawl, as a pattern, although help us to understand the spatial distribution but as a static phenomenon; in fact areas described as sprawled are typically part of a dynamic urban scene (Harvey and Clark 1965; Ewing 1997). The dynamics of sprawl process can be understood from the theoretical framework of urban growth process. Herold et al. (2005b) presents a hypothetical schema of urban growth process using a general conceptual representation (Fig. 1.2). According to Herold et al. (2005b), urban area expansion starts with a historical seed or *core* that grows and disperses to new individual development centres. This process of *diffusion* continues along a trajectory of organic growth and outward expansion. The continued spatial evolution transitions to the *coalescence* of the individual urban blobs. This phase transition initially includes development in the open space between the central urban core and peripheral centres. This conceptual growth pattern continues and the system progresses toward a saturated state. In Fig. 1.2, this 'final' agglomeration can be seen as an initial urban core for further urbanisation at a less detailed zoomed-out extent. In most traditional urbanisation-studies this 'scaling up' has been represented by changing the spatial extent of concentric rings around the central urban core.

The preceding framework suggests that some parts of an urban area may pass through a sprawl stage before eventually thickening so that they can no longer be characterised as sprawl. However, from this point of view what, when and where it

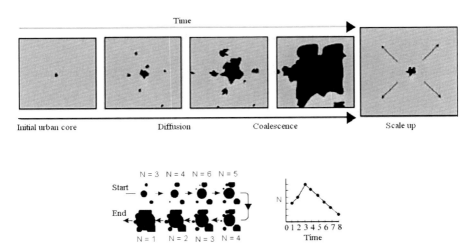

Fig. 1.2 Sequential frames of urban growth. The graph on the bottom-right shows N, number of agglomerations, through a sequence of time steps (Herold et al. 2005b)

can be characterised as sprawl, becomes ambiguous. Therefore, sprawl as a process without considering the pattern can not be characterised. Rather, it should be considered as a pattern in the light of multiple temporal process snapshots. 'In any event, measuring the respective dimensions of development patterns for an urban area at different times will reveal the process (or progress) of sprawl' (Galster et al. 2001).

Chapter 2
Causes and Consequences of Urban Growth and Sprawl

2.1 Introduction

An overall idea about urban growth and sprawl has been provided in Chap. 1. This chapter is aimed to list the causes and consequences of urban growth and sprawl. The causes that force growth in urban areas and the causes that are responsible for undesirable pattern or process of urban growth are also essentially important for the analysis of urban growth. The consequences or the impacts of urban growth, whether ill or good, are also necessary to be understood and evaluated towards achieving a sustainable urban growth.

Galster et al. (2001) argue that sprawl as a pattern or a process is to be distinguished from the causes that bring such a pattern about, or from the consequences of such patterns. This statement clearly says that analysis of pattern and process should be differentiated from the analysis of causes and consequences. Remote sensing data are more widely used for the analysis of pattern and process rather than causes or consequences. However, some of the researchers (e.g., Ewing 1994) argue that impacts of development present a specific development patterns as undesirable, not the patterns themselves. Therefore, whether a pattern is good or bad should be analysed from the perspective of its consequences. Causes are also similarly important to know the factors that are responsible to bring such pattern. Indeed remote sensing data are not enough to analyse the causes or consequences in many instances; one should have clear understanding of causes and consequences of urban growth and sprawl to encounter the associated problems.

2.2 Causes of Urban Growth and Sprawl

The causes of urban growth are quite similar with those of sprawl. In most of the instances they can not be discriminated since urban growth and sprawl are highly interlinked. However, it is important to realise that urban growth may be observed without the occurrence of sprawl, but sprawl must induce growth in urban area. Some of the causes, for example population growth, may result in coordinated compact growth or uncoordinated sprawled growth. Whether the growth is good or bad

Table 2.1 Causes of urban growth which may result in compact and/or sprawled growth

Causes of urban growth	Compact growth	Sprawled growth
Population growth	•	•
Independence of decision		•
Economic growth	•	•
Industrialisation	•	•
Speculation		•
Expectations of land appreciation		•
Land hunger attitude		•
Legal disputes		•
Physical geography		•
Development and property tax		•
Living and property cost		•
Lack of affordable housing		•
Demand of more living space	•	•
Public regulation		•
Transportation	•	•
Road width		•
Single-family home		•
Nucleus family	•	•
Credit and capital market		•
Government developmental policies		•
Lack of proper planning policies		•
Failure to enforce planning policies		•
Country-living desire		•
Housing investment		•
Large lot size		•

depends on its pattern, process, and consequences. There are also some of the causes that are especially responsible for sprawl; they can not result in a compact neighbourhood. For example, country-living desire—some people prefer to live in the rural countryside; this tendency always results in sprawl. Table 2.1 lists the causes of urban growth, and shows which of them may result in compact growth and which in sprawled growth.

The causes and catalysts of urban growth and sprawl, discussed by several researchers, can be summarised as presented in the following sections (for a general discussion one may refer Burchfield et al. 2006; Squires 2002; Harvey and Clark 1965).

2.2.1 Population Growth

The first and foremost reason of urban growth is increase in urban population. Rapid growth of urban areas is the result of two population growth factors: (1) natural increase in population, and (2) migration to urban areas. Natural population growth results from excess of births over deaths. Migration is defined as the long-term relocation of an individual, household or group to a new location outside the community of origin. In the recent time, the movement of people from rural to urban areas within the country (internal migration) is most significant. Although very insignificant comparing the movement of people within the country; international migration is also increasing. International migration includes labour migration, refugees and undocumented migrants. Both internal and international migrations contribute to urban growth.

Internal migration is often explained in terms of either *push factors*—conditions in the place of origin which are perceived by migrants as detrimental to their well-being or economic security, and *pull factors*—the circumstances in new places that attract individuals to move there. Examples of push factors include high unemployment and political persecution; examples of pull factors include job opportunities or better living facilities. Typically, a pull factor initiates migration that can be sustained by push and other factors that facilitate or make possible the change. For example, a farmer in rural area whose land has become unproductive because of drought (push factor) may decide to move to a nearby city where he perceives more job opportunities and possibilities for a better lifestyle (pull factor).

In general, cities are perceived as places where one could have a better life; because of better opportunities, higher salaries, better services, and better lifestyles. The perceived better conditions attract poor people from rural areas. People move into urban areas mainly to seek economic opportunities. In rural areas, often on small family farms, it is difficult to improve one's standard of living beyond basic sustenance. Farm living is dependent on unpredictable environmental conditions, and during of drought, flood or pestilence, survival becomes extremely problematic. Cities, in contrast, are known to be places where money, services and wealth are centralised. Cities are places where fortunes are made and where social mobility[1] is possible. Businesses that generate jobs and capitals are usually located in urban areas. Whether the source is trade or tourism, it is also through the cities that foreign money flows into a country. People living on a farm may wish to move to the city and try to make enough money to send back home to their struggling family.

In the cities, there are better basic services as well as other specialist services that are not found in rural areas. There are more job opportunities and a greater variety of jobs in the cities. Health is another major factor. People, especially the elderly are often forced to move to cities where there are doctors and hospitals that can cater for their health needs. Other factors include a greater variety of entertainment (restaurants, movie theatres, theme parks, etc.) and a better quality of education. Due to high populations, urban areas can also have much more diverse social communities allowing others to find people like them.

---

[1] Change in an individual's social class position (upward or downward) throughout the course of their life either between their own and their parents' social class (inter-generational mobility) or over the course of their working career (intra-generational mobility).

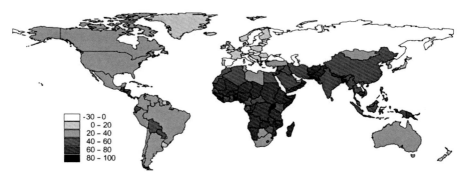

Fig. 2.1 Projected percentage increase in urban population 2000–2030 (United Nations 2002)

These conditions are heightened during times of change from a pre-industrial society to an industrial one. At this transition time many new commercial enterprises are made possible, thus creating new jobs in cities. It is also a result of industrialisation that farms become more mechanised, putting many farm labourers out of work. Developing nations are currently passing through the process of industrialisation. As a result, growth rate of urban population is very high in these countries comparing industrialised countries.

In industrialised countries the future growth of urban populations will be comparatively modest since their population growth rates are low and over 80% of their population already live in urban areas. In contrast, developing countries are in the middle of the transition process, when urban population growth rates are very high. According to the United Nations report (UNFPA 2007), the number and proportion of urban dwellers will continue to rise quickly (Fig. 2.1). Urban global population will grow to 4.9 billion by 2030. In comparison, the world's rural population is expected to decrease by some 28 million between 2005 and 2030. At the global level, all future population growth will thus be in towns and cities; most of which will be in developing countries. The urban population of Africa and Asia is expected to be doubled between 2000 and 2030.

This huge growth in urban population may force to cause uncontrolled urban growth resulting in sprawl. The rapid growth of cities strains their capacity to provide services such as energy, education, health care, transportation, sanitation, and physical security. Since governments have less revenue to spend on the basic upkeep of cities and the provision of services, cities become areas of massive sprawl and serious environmental problems.

2.2.2 Independence of Decision

The competitors (government and/or private) hold a variety of expectations about the future and a variety of development demands. Often these competitors can take decisions at their own to meet their future expectations and development demands.

This is especially true if the city lacks a master plan as a whole. This independence ultimately results in uncoordinated, uncontrolled and unplanned development (Harvey and Clark 1965).

2.2.3 Economic Growth

Expansion of economic base (such as higher per capita income, increase in number of working persons) creates demand for new housing or more housing space for individuals (Boyce 1963; Giuliano 1989; Bhatta 2009b). This also encourages many developers for rapid construction of new houses. Rapid development of housing and other urban infrastructure often produces a variety of discontinuous uncorrelated developments. Rapid development is also blamed owing to its lack of time for proper planning and coordination among developers, governments and proponents.

2.2.4 Industrialisation

Establishment of new industries in countryside increases impervious surfaces rapidly. Industry requires providing housing facilities to its workers in a large area that generally becomes larger than the industry itself. The transition process from agricultural to industrial employment demands more urban housing. Single-storey, low-density industrial parks surrounded by large parking lots are one of the main reasons of sprawl. There is no reason why light industrial and commercial land-uses cannot grow up instead of out, by adding more storeys instead of more hectares. Perhaps, industrial sprawl has happened because land at the urban edge is cheaper.

2.2.5 Speculation

Speculation about the future growth, future government policies and facilities (like transportation etc.) may cause premature growth without proper planning (Clawson 1962; Harvey and Clark 1965). Several political election manifestos may also encourage people speculating the direction and magnitude of future growth. Speculation is sometimes blamed for sprawl in that speculation produces withholding of land for development which is one reason of discontinuous development.

2.2.6 Expectations of Land Appreciation

Expectations of land appreciation at the urban fringe cause some landowners to withhold land from the market (Lessinger 1962; Ottensmann 1977). Expectations may vary, however, from landowner to landowner, as does the suitability of land for development. The result is a discontinuous pattern of development. The higher the rate of growth in a metropolitan area, the greater the expectations of land appreciation; as a result, more land will be withheld for future development.

2.2.7 Land Hunger Attitude

Many institutions and even individuals desire for the ownership of land. Often these lands left vacant within the core city area and makes infill policies unsuccessful (Harvey and Clark 1965). As a result the city grows outward leaving the undeveloped land within the city.

2.2.8 Legal Disputes

Legal disputes (e.g., ownership problem, subdivision problem, taxation problem, and tenant problem) often causes to left vacant spaces or single-storied buildings within the inner city space. This also causes outgrowth leaving the undeveloped land or single-storied buildings within the city.

2.2.9 Physical Geography

Sometimes the sprawl is caused because of unsuitable physical terrain (such as rugged terrain, wetlands, mineral lands, or water bodies, etc.) for continuous development (Fig. 2.2). This often creates leap-frog development sprawl (Harvey and Clark 1965; Barnes et al. 2001). Important to mention that in many instances these problems cannot be overcome and therefore should be overlooked.

Fig. 2.2 Unsuitable physical terrain prohibits continuous development

2.2.10 Development and Property Tax

Generally, the costs involved in development of community-infrastructure and public services are higher in the countryside rather than the core city (refer Sect. 2.3.1). The maintenance costs of public services are also higher in the countryside. Therefore, the development and property tax should be higher at the periphery of the city. However, generally these taxes are independent of location and even in many instances these taxes are lower in the periphery comparing the core city. The problem is that local tax systems usually require developers to pay only a fraction of the community-infrastructure and public-service costs associated with their projects, which makes development look artificially cheap and encourages urban expansion (Brueckner and Kim 2003). Underpricing of urban infrastructure encourages excessive spatial growth of cities, as shown by Brueckner (1997).

2.2.11 Living and Property Cost

Generally living cost and property cost is higher in the main city area than the countryside. This encourages countryside development. Harvey and Clark (1965) say 'at the time of sprawl occurred, the cost was not prohibitive to the settler, (rather) it provided a housing opportunity economically satisfactory relative to other alternatives'. Generally majority of urban residents seek to settle within the core city, but lower living and property cost attract them to the countryside.

2.2.12 Lack of Affordable Housing

It is similar to *living and property cost* and another reason of urban sprawl. Affordable housing is a term used to describe dwelling units whose total housing costs are deemed 'affordable' to those that have a median household income.[2] A common measure of community-wide affordability is the number of homes that a household with a certain percentage of median income can afford. For example, in a perfectly balanced housing market, the median household (and the half of the households which are wealthier) could officially afford the median housing option, while those poorer than the median income could not afford the median home. 50% affordability for the median home indicates a balanced market. Lack of affordable housing within the city forces people to set their residences in the countryside.

---

[2] The median household income (or median income) is commonly used to provide data about geographic areas and divides households into two equal segments with the first half of households earning less than the median household income and the other half earning more. The median income is considered by many statisticians to be a better indicator than the average household income as it is not dramatically affected by unusually high or low values.

2.2.13 Demand of More Living Space

In many developing countries, residents of the core city lack sufficient living space. This encourages countryside development for more living space. People can buy more living space in the countryside than in the inner city, since the cost of property is less in the countryside. However, consumption of more living space not always causes sprawl. Population density is a major concern in this issue. Cities in developing countries are three times denser than the cities in developed countries (Acioly and Davidson 1996). Therefore, higher per capita consumption of built-up area (or living space) is desired in many instances. In such cases, higher per capita consumption of living space may indicate better and extended living facilities within the confines of compact urban growth. However, if the demand of more living space forces rapid low-density development in the countryside then it must be an indication of sprawl.

2.2.14 Public Regulation

Generally outside of the main city is lesser controlled and loosely regulated. As a result, many developers and individuals find these places more suitable for new construction (Harvey and Clark 1965). Loosely regulated public regulations also fail to control the new construction in a compact and sustainable manner, and in many instances developers do not bother about the government planning policies.

2.2.15 Transportation

Transportation routes open the access of city to the countryside and responsible for linear branch development (Fig. 2.3). The construction of expressways and highways cause both congestion in the city and rapid outgrowth (Harvey and Clark 1965). Roads are commonly considered in modelling and forecasting urban sprawl (Cheng and Masser 2003; Yang and Lo 2003), because they are a major catalyst of sprawl. Important to realise that transportation facilities are essential to cities and its neighbourhoods. Development of urban economy and thereby job opportunities are directly dependent on the transportation facilities. Therefore, transportation facilities can never be suppressed; rather initiatives to impede linear branch development by means of government policies and regulations should be practiced.

2.2.16 Road Width

Governments do not allow construction of high-rise buildings if the site can not be easily accessed. Narrow roads within the city area restrict construction of high-rise buildings resulting in waste of vertical space. This wastage of vertical space transformed into horizontal growth. This is a common problem to very old cities in

Fig. 2.3 Construction of roads encourages linear branch sprawl

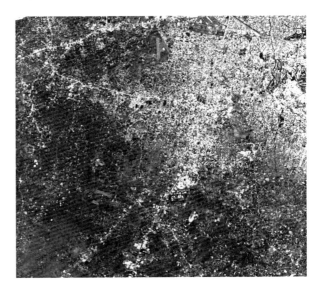

many developing countries where past planners failed to visualise the future needs and did not plan wider roads. Recent road-widening policies that are taken in many developing countries have failed owing to their economic (huge money is required to compensate the road-side house owners) and political constrains.

2.2.17 Single-Family Home

In many instances, individuals built a single-family home (rather than multi-family high-rise building) (Fig. 2.4). This also wastes vertical space significantly resulting in horizontal growth. Single-family residences increase the size of a city in multiple magnitudes.

Fig. 2.4 Single-family homes waste vertical space

2.2.18 Nucleus Family

Commonly, percapita consumption of carpet area in nucleus family is higher than the joint family. For example, a common dining space is shared by all the family members in a joint family. Transition from joint family to nucleus family also creates demand of new housing for individuals.

2.2.19 Credit and Capital Market

Credit and loan facility, low interest rate, etc. are also responsible for rapid urban growth in advance. In this sense, people can buy homes before achieving the financial capability. Therefore, the growth will occur in advance than actually supposed to be.

2.2.20 Government Developmental Policies

Restrictive land-use policies in one political jurisdiction may lead development to 'jump' to one that is favourably disposed toward development or is less able to prevent or control it (Barnes et al. 2001). Often dissimilarities in development regulations, land-use policies, and urban services among the neighbouring municipalities (or local governments) may cause discontinuous development.

2.2.21 Lack of Proper Planning Policies

Lack of consistent and well-experimented planning policies may also cause urban sprawl. A city may be planned with exclusive zoning policies; this means separation of residential, commercial, industrial, office, institutional, or other land uses. Completely separate zoning created isolated islands of each type of development. In most cases, the automobile had become a requirement for transportation between vast fields of residentially zoned housing and the separate commercial and office strips, creating issues of automobile dependency and more fossil fuel consumption and thereby pollution. A mixed land-use policy is preferred to fight against sprawl.

2.2.22 Failure to Enforce Planning Policies

Having a proper planning policy is not enough, rather its successful implementation and enforcement is more important. Unsuccessful enforcement of land-use plans is one of the reasons of sprawl in developing countries, since the enforcement is often corrupt and intermittent in these countries.

2.2.23 Country-Living Desire

Residents of countryside are often former urbanites who desire the solitude and perceived amenities of country-living as rural retreats. Despite traffic congestion and long commutes to work, moving to the suburbs remains a goal for many city residents who perceive quality of life in the suburbs as better. Unless this perception changes and the conditions of urban life improve, sprawl development will continue as the flight from cities to suburbs continues (Barnes et al. 2001).

2.2.24 Housing Investment

Often urbanites purchase second homes in the countryside as future investments (Barnes et al. 2001). This encourages the developers for construction at the countryside in advance. These homes often left vacant but the government is forced to maintain urban facilities and services in a low-density area. Low interest rate and high housing demand make the countryside-housing investment more attractive.

2.2.25 Large Lot Size

Large lot (or plot) size is another reason of sprawl. Large-lot residents utilise a portion of their land for construction purposes leaving other portions as non-developed (Fig. 2.5). Although this problem is mainly associated with developed countries; however, also in the developing countries, residents in the countryside generally prefer to have a large individual lot.

Fig. 2.5 Large-lot residents utilise a portion of their land for construction

2.3 Consequences of Urban Growth and Sprawl

Consequences of urban growth may have both positive and negative impacts; however, negative impacts are generally more highlighted because this growth is often uncontrolled or uncoordinated and therefore the negative impacts override the positive sides. Positive implications of urban growth include higher economic production, opportunities for the underemployed and unemployed, better life because of better opportunities and better services, and better lifestyles. Urban growth can extend better basic services (such as transportation, sewer, and water) as well as other specialist services (such as better educational facilities, health care facilities) to more peoples. However, in many instances, urban growth is uncontrolled and uncoordinated resulting in sprawl. As a result, the upside impacts vanish inviting the downsides.

In the developed countries, during the nineteenth and early twentieth centuries, urbanisation resulted from and contributed to industrialisation. New job opportunities in the cities motivated the mass movement of surplus population away from the villages. At the same time, migrants provided cheap, plentiful labour for the emerging factories. Currently, due to movements such as globalisation, the circumstances are similar in developing countries. The concentration of investments in cities attracts large number of migrants looking for employment, thereby creating a large surplus labour force, which keeps wages low. This situation is attractive to foreign investment companies from developed countries who can produce goods for far less than if the goods were produced where wages are higher. Thus, one might wonder if urban poverty serves a distinct function for the benefit of global capital.

Developed and developing countries of the world differ not only in the number of people living in cities, but also in the way in which urbanisation is occurring. In many megacities of developing world, urban sprawl is a common problem and a substantial amount of city dwellers live in slums within the city or in urban periphery in poverty and degraded environment (Fig. 2.6). These high-density settlements are often highly polluted owing to the lack of urban services, including running water, sewer, trash pickup, electricity or paved roads. Nevertheless, cities provide poor people with more opportunities and greater access to resources to transform their situation than rural areas.

Fig. 2.6 Housing of poor urban people

2.3 Consequences of Urban Growth and Sprawl

One of the major effects of rapid urban growth is sprawl that increases traffic, saps local resources, and destroys open space. Urban sprawl is responsible for changes in the physical environment, and in the form and spatial structure of cities. In many countries including the developed countries like United States, poorly planned urban development is threatening the environment, health, and quality of life. In communities across the world, sprawl is taking a serious toll.

Evidence of the environmental impacts of sprawl continues to mount. Kirtland et al. (1994) report that the impact of urban land on environmental quality is much larger than its spatial extent would imply. The consequences and significance of sprawl, good or ill, are evaluated based on its socioeconomic and environmental impacts. Often these are overlapping or one may have several indirect impacts. However, major consequences of urban sprawl can be summarised as follows.

2.3.1 Inflated Infrastructure and Public Service Costs

Sprawl is usually accepted as being inordinately costly to its occupants and to society (Harvey and Clark 1965). Sprawl is blamed due to its environmental cost and economic cost (Buiton 1994). Cities have experienced an increase in demand for public services and for the maintenance and improvement of urban infrastructures (Barnes et al. 2001) such as fire-service stations, police stations, schools, hospitals, roads, water mains, and sewers in the countryside. Sprawl requires more infrastructures, since it takes more roads, pipes, cables and wires to service these low-density areas compared to more compact developments with the same number of households. Other services such as waste and recyclables collection, mail delivery and street cleaning are more costly in low-density developments, while public transit is impractical because the rider density needed to support a transit service is not there.

The Costs of Sprawl and other studies have shown that development of neighbourhood infrastructure becomes less costly on a per-unit basis as density rises (refer Priest et al. 1977; Frank 1989). As long as developers are responsible for the full costs of neighbourhood infrastructure, and pass such costs on to homebuyers and other end-users of land, lower-density development patterns will meet the test of economic efficiency (at least with respect to infrastructure costs). Where inefficiency is more likely to arise is in the provision of community-level infrastructure. Inefficiency may also arise in the operation and maintenance of infrastructure, and in the provision of public services. Because people are more dispersed and no longer residing in centralised cities, the costs of community infrastructure and public services in suburban areas increases (Brueckner 2000; Heimlich and Anderson 2001; Pedersen et al. 1999; Wasserman 2000). These costs tend to be financed with local taxes or user fees that are generally independent of location, causing remote development to be subsidised.

It may be mentioned that from the standpoint of community-level infrastructure, costs do not vary so much with residential density but with the degree of clustering and/or proximity to existing development (HCPC 1967; Stone 1973; RERC

1974; Downing and Gustely 1977; Peiser 1984). So, too, the costs of public services (Archer 1973; RERC 1974; Downing and Gustely 1977; Peiser 1984).

2.3.2 Energy Inefficiency

Higher densities mean shorter trips but more congestion. Newman and Kenworthy (1988) find that the former effect overwhelms the latter. Even though vehicles are not as fuel-efficient in dense areas owing to traffic congestion, fuel consumption per capita is still substantially less in dense areas because people drive so much less. Urban sprawl causes more travel from the suburbia to the central city and thus more fuel consumption. Furthermore, it also causes traffic congestion. More cars on the roads driving greater distances are a recipe for traffic gridlock resulting in more fuel consumption.

With electricity, there is a cost associated with extending and maintaining the service delivery system, as with water, but there also is a loss in the commodity being delivered. The farther from the generator, the more power is lost in distribution.

2.3.3 Disparity in Wealth

There is marked spatial disparity in wealth between cities and suburbs; and sprawled land development patterns make establishing and using mass transit systems difficult (Benfield et al. 1999; Kunstler 1993; Mitchell 2001; Stoel 1999). Sprawl is also implicated in a host of economic and social issues related to the deterioration of urban communities and the quality of life in suburbia (Wilson et al. 2003). In many cases private utility systems serving the main segment of the settled area cannot be expanded for technical and financial reasons. Urban sprawl often occurs in peripheral areas without the discipline of proper planning and zoning; as a result, it blocks the ways of future possible quality services.

2.3.4 Impacts on Wildlife and Ecosystem

In areas where sprawl is not controlled, the concentration of human presence in residential and industrial settings may lead to an alteration of ecosystems patterns and processes (Grimm et al. 2000). Development associated with sprawl not only decreases the amount of forest area (Macie and Moll 1989; MacDonald and Rudel 2005), farmland (Harvey and Clark 1965), woodland (Hedblom and Soderstrom 2008), and open space but also breaks up what is left into small chunks that disrupt ecosystems and fragment habitats (Lassila 1999; McArthur and Wilson 1967; O'Connor et al. 1990). The reach of urban sprawl into rural natural areas such as woodlands and wetlands ranks as one of the primary forms of wildlife habitat loss. Roads, power lines, subdivisions and pipelines often cut through natural areas, thereby fragmenting wildlife habitat and altering wildlife movement patterns

2.3 Consequences of Urban Growth and Sprawl

Fig. 2.7 Fragmentation of wildlife habitat

(Fig. 2.7). The fragmentation of a large forest into smaller patches disrupts ecological processes and reduces the availability of habitat for some species. Some forest fragments are too small to maintain viable breeding populations of certain wildlife species.

2.3.5 Loss of Farmland

Urbanisation generally, and sprawl in particular, contribute to loss of farmlands and open spaces (Berry and Plaut 1978; Fischel 1982; Nelson 1990; Zhang et al. 2007). Urban growth, only in the United States, is predicted to consume 7 million acres of farmland, 7 million acres of environmentally sensitive land, and 5 million acres of other lands during the period 2000–2025 (Burchell et al. 2005). This case is enough to visualise the world scenario.

Provincial tax and land-use policies combine to create financial pressures that propel farmers to sell land to speculators. Low prices of farm commodity in global markets often mean it is far more profitable in the long term for farmers to sell their land than to continue farming it. In addition, thousands of relatively small parcels of farmland are being severed off to create rural residential development. Collectively, these small lots contribute to the loss of hundreds of hectares of productive agricultural land per year.

The loss of agricultural land to urban sprawl means not only the loss of fresh local food sources but also the loss of habitat and species diversity, since farms include plant and animal habitat in woodlots and hedgerows. The presence of farms on the rural landscape provides benefits such as greenspace, rural economic stability, and preservation of the traditional rural lifestyle.

2.3.6 Increase in Temperature

Positive correlation between land surface temperature and impervious surface clearly indicates temperature increase in the sprawled area (Weng et al. 2007; Wang et al. 2003). On warm days, urban areas can be 6–8°F (3.5–4.5°C) warmer than surrounding areas, an effect known as an *urban heat island* (Frumkin 2002) (Fig. 2.8). The heat island effect is caused by two factors. First, dark surfaces such as roadways and rooftops efficiently absorb heat from sunlight and reradiate it as thermal infrared radiation; these surfaces can reach temperatures of 50–70°F (28–39°C) higher than surrounding air. Second, urban areas are relatively devoid of vegetation, especially trees; that would provide shade and cool the air through *evapotranspiration*.[3] As cities sprawl outward, the heat island effect expands, both in geographic extent and in intensity. This is especially true if the pattern of development features extensive tree-cutting and road construction.

Furthermore, dispersed metropolitan expansion involves a positive feedback loop that may aggravate the heat island effect. Sprawling metropolitan areas, with greater travel distances, generate a large amount of automobile travel. This, in turn, results in more fuel combustion, with more production of carbon dioxide, and consequent contributions to global climate change. Global climate change, in turn, may intensify the heat island effect in metropolitan areas. Thus, not only does the morphology of metropolitan areas contribute to warming, but so may the greenhouse gas production that results from increased driving.

The number of habitants is a decisive factor conditioning the occurrence of urban heat island. Figure 2.9 shows increased city size (represented by circles) with increasing number of habitants is responsible for increasing urban temperature.

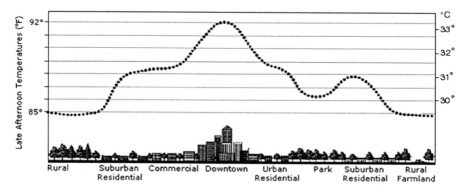

Fig. 2.8 An urban heat island profile (Klinenberg 2002)

---

[3] Evapotranspiration is a term used to describe the sum of evaporation and plant transpiration from the earth's land surface to atmosphere.

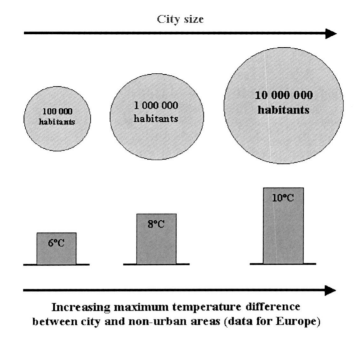

Fig. 2.9 Increased city size and number of habitants cause increase in temperature (Source: http://www.atmosphere.mpg.de)

2.3.7 Poor Air Quality

Sprawl is cited as a factor of air pollution (Stone 2008), since the car-dependent lifestyle imposed by sprawl leads to increases in fossil fuel consumption and emissions of greenhouse gases (Stoel 1999). Urban sprawl contributes to poorer air quality by encouraging more automobile use, thereby adding more air pollutants such as carbon monoxide, carbon dioxide, ground-level ozone, sulphur dioxide, nitrogen oxides, volatile organic carbons, and microscopic particles (Frumkin 2002). These pollutants can inhibit plant growth, create smog and acid rain, contribute to global warming, and cause serious human health problems. Apparently it seems that low-density urban growth or sprawl can provide better environmental condition and fresh air, but Kahn and Schwartz (2008) found that urban air pollution progress despite sprawl.

Increased temperature in urban areas also has indirect effects on air pollution. As the temperature rises, so does the demand for energy to power fans, air coolers, water coolers, and air conditioners; requiring power plants to increase their output. The majority of power plants burn fossil fuels, so increased demand of power in summer results in higher emissions of the pollutants they generate, including carbon dioxide, particulate matter, sulphur oxides, nitrogen oxides, and air toxics. Furthermore, ozone formation from its precursors, nitrogen oxides and hydrocarbons, is enhanced by heat (Frumkin 2002).

2.3.8 Impacts on Water Quality and Quantity

Sprawl also has serious impacts on water quality and quantity. With miles of roads, parking lots and houses having paved over the countryside, rainwater and snowmelt are unable to soak into the ground and replenish the groundwater aquifers.

Urban growth and sprawl lead to an increasing imperviousness, which in turn induces more total runoff volume. So urban areas located in flood-prone areas are exposed to increased flood hazard, including inundation and erosion (Jacquin et al. 2008). As new development continues in the periphery of the existing urban landscape, the public, the government, planners and insurance companies are more and more concerned by flooding disasters and increasing damages (Wisner et al. 2004; Jacquin et al. 2008).

In the urban area, water runs off into storm sewers and ultimately into rivers and lakes. Extra water during heavy rain can dramatically increase the rate of flow through wetlands and rivers, stripping vegetation and destroying habitats along riverbanks. It can also cause damaging floods downstream and lead to an increase in water pollution from runoff contaminated with lawn and garden chemicals, motor oil and road salt. Widely dispersed development requires more pavements that cause more urban runoff that pollutes waterways (Lassila 1999; Wasserman 2000). These pollutants can be absorbed by humans when they eat contaminated fish from affected water-bodies and when they drink from contaminated surface water or groundwater sources.

In addition, heavy rainstorms occurring in cities and towns with inadequate systems for managing stormwater can cause untreated human sewage to enter waterways (*combined sewer overflow*).

2.3.9 Impacts on Public and Social Health

One of the original motivations for migration to the suburbs was access to nature. People generally prefer to live with trees, birds, and flowers; and these are more accessible in the suburbs than in denser urban areas. Moreover, contact with nature may offer benefits beyond the purely aesthetic; it may benefit both mental and physical health. In addition, the sense of escaping from the turmoil of urban life to the suburbs, the feeling of peaceful refuge, may be soothing and restorative to some people. In these respects, there may be health benefits to suburban lifestyles (Frumkin 2002). However, sprawl is generally blamed for its negative impacts on public health (refer Frumkin 2002; Savitch 2003; Yanos 2007; Sturm and Cohen 2004).

One of the cardinal features of sprawl is driving, reflecting a well-established, close relationship between lower density development and more automobile travel.

2.3 Consequences of Urban Growth and Sprawl

Automobile use offers extraordinary personal mobility and independence. However, it is also associated with health hazards, including air pollution, motor vehicle crashes, and pedestrian injuries and fatalities (Frumkin 2002). Air pollution causes severe breathing problems, skin diseases, and other health problems. The effects of air pollution on the health of human and other living species are perhaps known to everyone.

Sprawl is blamed for driving out local downtown commerce by attracting consumers to larger, regional malls and restaurants (Pedersen et al. 1999). Sprawl results waste in time of passing vacant land enroute from central city to the sprawled suburb (Harvey and Clark 1965), giving rise to more traffic congestion (Brueckner 2000; Ewing 1997; Pedersen et al. 1999; Wasserman 2000) and reduced social interaction. Since sprawl is so car-dependent, walking or cycling opportunities (and the chances they bring for social interaction) diminish, while driving distances tend to lengthen dramatically. Long commutes to and from work heighten psychological stress. As people spend more time on more crowded roads, an increase in these psychological health-outcomes might be expected. One possible indicator of such problems is road rage, defined as 'events in which an angry or impatient driver tries to kill or injure another driver after a traffic dispute' (Rathbone and Huckabee 1999). Longer travel-time also reduces time available for work, leisure, and family (Wilson et al. 2003). Families who can not afford housing to live within the city may suffer from distress that may cause negative impacts on a community's overall health.

Rates of automobile fatalities and injuries per driver and per mile driven have fallen thanks to safer cars and roads, seat belt use, laws that discourage drunk driving, and other measures, but the absolute toll of automobile crashes remains high. The relationship between sprawl and motor vehicle crashes is complex. At the simplest level, more driving means greater exposure to the dangers of the road, translating to a higher probability of a motor vehicle crash. Suburban roads may be a particular hazard, especially major commercial thoroughfares and 'feeder' roads that combine high speed, high traffic volume, and frequent 'curb cuts' for drivers to use in entering and exiting stores and other destinations (Frumkin 2002). The most dangerous stretches of road were those built in the style that typifies sprawl: multiple lanes, high speeds, no sidewalks, long distances between intersections or crosswalks, and roadways lined with large commercial establishments and apartments blocks. Walking offers important public health benefits, but safe and attractive sidewalks and footpaths are needed to attract walkers and assure their safety that is often suffered by sprawled development.

Urban areas are warmer than rural. Heat is of concern because it is a health hazard. Relatively benign disorders include heat syncope, or fainting; heat edema, or swelling, usually of dependent parts such as the legs; and heat tetany, a result of heat-induced hyperventilation. Other effects include heat cramps, heat exhaustion vomiting, weakness, and mental status changes. The most serious of the acute heat-related conditions is heat stroke. Frumkin (2002) discussed these urban health issues in detail.

From the perspective of social health, low-density development is blamed for reducing social interaction and threatening the ways that people live together (Ewing

1997; Putnam 2000). Residents may also lose their sense of community as their town's population swells dramatically.

2.3.10 Other Impacts

Exurban development can place additional burdens on rural economic/land-use activities such as forestry, mining, and farming, since the values of exurbanites may clash with those of traditional users regarding the most suitable uses of rural lands.

Urban sprawl, a potential manifestation of development, has its negative impacts in coastal regions also, where beach-oriented tourism and amenity-driven population growth and land development are prominent (Crawford 2007).

Sprawl also includes aesthetic impacts such as more ugly and monotonous suburban landscapes. For other several indirect impacts of sprawl please refer Barnes et al. (2001) and Squires (2002).

Chapter 3
Towards Sustainable Development and Smart Growth

3.1 Introduction

This chapter provides basic concepts on sustainable development and smart growth. Detailed discussion on these issues is not within the scope of this book since it is mainly focused on the analysis of urban growth pattern and process. However, a basic understanding on sustainable development and smart growth will help to analyse the urban growth in terms of its acceptability of growth pattern/process or planning for the future. This chapter also documents several studies that are aimed to restrict the freedom of urban growth towards achieving the goals of smart and sustainable urban growth.

3.2 Sustainable Development

The idea of sustainability dates back more than 40 years, to the new mandate adopted by the International Union for Conservation of Nature in 1969 (IUCN 2006). It was a key theme of the United Nations Conference on the Human Environment in Stockholm in 1972. The concept of sustainability was coined explicitly to suggest that it was possible to achieve economic growth and industrialisation without environmental damage. At the start of the twenty-first century, the problem of global sustainability is widely recognised by world leaders, and a common topic of discussion by among various sections of the society—journalists, scientists, teachers, students and citizens in many parts of the world. The World Summit on Sustainable Development (WSSD 2002) confirmed that the first decade of the new century, at least, would be one of reflection about the demands placed by humankind on the biosphere.

Sustainable development is defined as, 'development that meets the needs of the present without compromising the ability of the future generations to meet their own needs' (WCED 1987). Sustainable development is a pattern of resource use that aims to meet human needs while preserving the environment so that these needs can be met not only in the present, but also for future generations. In order to sustain a development, the supply and quality of major consumables and inputs to our

Fig. 3.1 Schematic model of sustainability (after IUCN 2006)

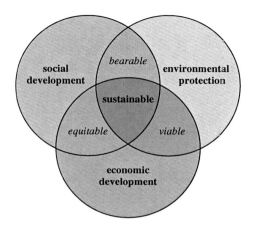

daily lives and economic production—such as air, water, energy, food, raw materials, land, and the natural environment needs to be taken care of. Land is essential because our food and raw materials originate from them and is a habitat for flora and fauna. Similar to other resources it is a scarce commodity. Any disturbance to this resource by way of change in land-use, e.g. conversion of forestland or agricultural land into built-up, is irreversible. The use of land unsuitable for development may be unsustainable for the natural environment as well as to the humans.

Sustainable development does not focus solely on environmental issues. The *United Nations 2005 World Summit Outcome Document* refers to the 'interdependent and mutually reinforcing pillars' of sustainable development as *economic development*, *social development*, and *environmental protection* (United Nations 2005b) (Fig. 3.1). The United Nations' *Millennium Development Goals Report* (United Nations 2009) recommends eight major goals with 16 targets. This report not only emphasises the environmental sustainability but also economic and social issues. According to Hasna (2007), sustainability is a process which tells of a development of all aspects of human life affecting sustenance. It means resolving the conflict between the various competing goals, and involves the simultaneous pursuit of *economic prosperity, environmental quality* and *social equity*. Hence it is a continually evolving process; the 'journey' (the process of achieving sustainability) is of course vitally important, but only as a means of getting to the destination (the desired future state). However, the 'destination' of sustainability is not a fixed place in the normal sense that we understand destination. Instead, it is a set of wishful characteristics of a future system. Sustainable development brings us to a destination that is bearable, equitable, and viable. The concepts of sustainable development are well documented in WCED (1987).

Important to mention, sustainable development does not mean *green development*. Green development prioritises environmental sustainability over economic and social development. Proponents of sustainable development argue that it provides a context in which cutting edge green development is unattainable to improve the overall sustainability. For example, a cutting edge treatment plant with extremely high maintenance costs may not be sustainable in regions of the world with fewer financial resources. An environmentally ideal plant that requires a huge maintenance cost is obviously less sustainable than one that is maintainable by the community, even if it is somewhat less effective from an environmental standpoint.

In 2004, the *Aalborg Commitments* initiative was launched to strengthen efforts towards urban sustainability in Europe (ICLEI 2004). The Aalborg Commitments are criteria and conditions that cities and towns can voluntarily undertake to abide by in order to translate a common vision of sustainable development into tangible sustainability targets and actions at the local level. At the heart of the Aalborg Commitments there are 50 sustainability criteria (consisting of ten topical groups each containing five sustainability criteria) that serve as guidelines for the planning, implementation and evaluation of sustainable urban development. A detailed discussion on *Aalborg Commitments* for urban sustainability can be found in Zilans and Abolina (2009).

With the increasing acceptance of sustainable development as a guiding concept, researchers have focused renewed attention on matters of urban form that trace back to the start of the modern planning and urban studies (Howard 1898; Burgess 1925; Hoyt 1939; Harris and Ullman 1945; Conzen 2001). A growing body of literature looks to a 'good city form' or 'sustainable urban form' to enhance economic vitality and social equity, and reduce the deterioration of the environment (Breheny 1992; De Roo and Miller 2000). In the last few years, numerous sustainability studies have been commissioned by government agencies that have led to the recommendation of suites of indicators to assess a host of related socio-economic and environmental issues. Many of these indicators have been described in concept only and a major challenge for science remains in translating descriptive indicators into quantifiable measures. How to measure the degree of sustainability is still to be explored and invented. In this regard, Zhang and Guindon (2006) may be referred who have proposed several methodologies for sustainability indicator quantification from remote sensing data.

3.3 Smart Growth

In the discipline of urban growth, a new word has emerged—*smart growth*, to attain the goals of sustainability (Downs 2001; Gillham and Maclean 2001; Lindstrom and Bartling 2003, Smart Growth Network 2003; Burchell et al. 2005; English 1999; Tregoning et al. 2002). The term 'smart growth' has been widely adopted to characterise compact patterns of development that do not embody the negative characteristics of sprawl (Danielsen et al. 1999; Hasse 2004). Smart growth is an urban planning and transportation theory that concentrates growth in the centre of a city to avoid urban sprawl; and advocates compact, transit-oriented, walkable, bicycle-friendly land-use, including neighbourhood schools, complete streets, and mixed-use development with a range of housing choices. Smart growth is linked to

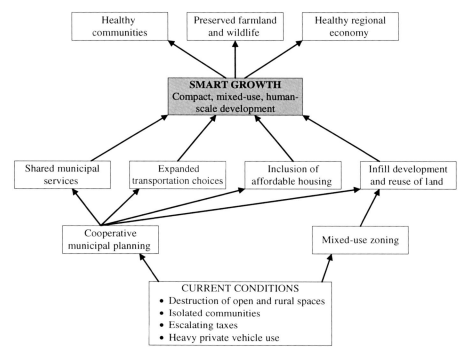

Fig. 3.2 A simple conceptual model of smart growth

the lofty ideals of a *new-urbanism* [1] and *neo-traditional town planning* (Brown and Bonifay 2001). Figure 3.2 presents a simplified conceptual model of smart growth.

Smart growth programme values long-range regional considerations of sustainability over a short-term focus. It balances the competing interests of the environment, the economy, and quality of life. Ostensibly, smart growth achieves this goal by emphasising the overall liveability of the locality and the importance of sense of place (English 1999). Smart growth mitigates the negative externalities of uncontrolled growth by placing controls on it. Because of new controls, the overall quality of economic development improves, the quality of life indicators rises, and the whole local system becomes inherently more sustainable (Calavita and Caves 1994; English 1999; Weitz 1999).

The goals of smart growth are to achieve a unique sense of community and place; expand the range of transportation, employment, and housing choices; equitably distribute the costs and benefits of development; preserve and enhance natural and cultural resources; and promote public health. However, smart growth principles are also sometimes criticised, due to difficulties in implementation because

---

[1] The Congress for the New Urbanism (www.cnu.org) is the leading organisation for promoting the concept of new-urbanism.

3.3 Smart Growth

they conflict with the tradition of low-density urban development (Downs 2001). Additionally, some smart growth policies have been shown to encourage sprawl (Irwin and Bockstael 2004).

In recent years, smart growth messages coming from environmentalists and government officials that recommend various techniques to reduce sprawl. However, very few local decision-makers have an instinctual feel for what these terms mean, let alone how they are translated into the landscape around them (Wilson et al. 2003). Smart growth programs often involve a package of tools such as mixed-use zoning, comprehensive plans, subdivision regulations, development fees, exactions, and infrastructure investments, applied together with high-density development (Nelson *et al.* 2002). Hasse (2004) points out that urban growth following the principles of smart growth – e.g., pedestrian-friendly development, multi-nodal transportation coordination, and urban redevelopment – holds the potential to lessen the environmental impacts and social costs of sprawling development. Staley and Gilroy (2004) emphasise increase in housing affordability and diversity as the core principle of most smart growth policies, stressing that low-density residential and commercial development reduces the overall quality of urban life through reliance on the automobile, whereas compact, higher density land use patterns improve the quality of life through a pedestrian lifestyle and provide a wide range of housing choice. Downs (2005) states that successful implementation of smart growth policies requires adopting policies that contradict long-established traditions such as low-density living patterns. Growth is 'smart growth', to the extent that it includes the elements discussed in the following sections (USEPA 2009; SCN 2009).

Smart growth policies have also been criticised by many researchers arguing that urban sprawl generates private benefits and that an important cost of smart growth is an increase in the price of housing (Segal and Srinivasan 1985; Conte 2000; Gordon and Richardson 2000). Private benefits from sprawl can arise from the satisfaction of consumer preferences for more socioeconomically segregated communities that are less densely settled and may be able to offer lower housing prices (Wassmer and Baass 2006). Further, in support of sprawl, Glaesar and Kahn (2004) pointed to cheaper and larger homes as a benefit. Burchell et al. (2000a,b) note that land further from the center of a metropolitan area is less expensive, resulting in cheaper housing. In their response to critics of sprawling land use patterns in the United States, Gordon and Richard (2000) emphasise that desires for larger houses and lot sizes are more likely to be met in outlying areas. Kahn (2001) noted that sprawl can increase housing affordability in central cities and suburbs. Levine (1999) shows that growth management measures in California have accelerated the movement of minorities and the poor from central cities.

3.3.1 Compact Neighbourhoods

Compact liveable urban neighbourhoods attract more people and business capitals. Creating such neighbourhoods is a critical element of reducing urban sprawl and protecting the climate. The concept of compact neighbourhood includes adopting redevelopment strategies and zoning policies that channel housing and job growth into urban centres and neighbourhood business districts, to create compact, walkable, and bike- and transit-friendly hubs. This sometimes requires local governmental bodies to implement policies that allow increased height and density towards downtown. Other topics that fall under this concept are as follows:

- *Mixed-use development*—the practice of allowing more than one type of use in a building or set of buildings. In planning terms, this can mean some combination of residential, commercial, industrial, office, institutional, or other land-uses. This can reduce the automobile dependency of completely separate zoning policies.
- *Inclusion of affordable housing*—this allows more people to live within the city. In addition to the distress it causes families who cannot easily find a place to live, lack of affordable housing is considered by many urban planners to have negative effects on a community's overall health. For example, lack of affordable housing can make low-cost labour scarcer, and increase demands on transportation systems (as workers travel longer distances) (Pollard and Stanley 2007).
- *Restrictions or limitations on suburban design forms*—for example, discouraging detached houses on individual lots, strip malls[2] and surface parking lots.
- *Inclusion of parks and recreation areas*—for better environment and society.

3.3.2 Transit-Oriented Development

Transit-oriented development is a residential or commercial area designed to maximise access to public transport, and mixed-use/compact neighbourhoods tend to use transit at all times of the day. Availability and easy access of public transport reduces the use of vehicles for individuals. Many cities striving to implement better transit-oriented strategies seek to secure funding to create new public transportation infrastructure and improve existing services. Other measures might include regional cooperation to increase efficiency and expand services, and moving buses and trains more frequently through high-use areas. In short, transit-oriented developments are coupling a multi-modal approach to transportation with supportive development patterns, to create a variety of transportation options.

---

[2] Strip mall is an open area shopping center where the stores are arranged in a row, with a sidewalk in front. Strip malls are typically developed as a unit and have large parking lots in front.

3.3.3 Pedestrian- and Bicycle-Friendly Design

Biking and walking instead of driving can reduce emissions, save money on fuel and maintenance, and foster a healthier population. Pedestrian- and bicycle-friendly development includes bicycle lanes on main streets, an urban bicycle-trail system, bicycle parking, pedestrian crossings, and associated master plans. Important to realise, bicycles/bikes can not serve the purposes of main transportation. The goal of transportation is to transfer of maximum possible people or goods at a least possible time. Bicycles/bikes violate this basic assumption of transportation. Bicycles are preferred for shorter trips; and a separate lane for bicycles serve this purpose by reducing traffic congestion on the main streets and thereby pollution.

3.3.4 Others Elements

Other considerations of smart growth include the following:

- Preserving open space, farmland, critical ecological habitats, and natural beauty.
- Reusing of land.
- Foster distinctive, attractive communities with a strong sense of place.
- Protecting water and air quality.
- Transparent, predictable, fair and cost-effective rules for development.
- Taking advantage of compact building design.
- Historic preservation that refers to a professional endeavour that seeks to preserve, conserve and protect buildings, objects, landscapes or other artefacts of historic significance.
- Expansion around already developed areas. Development around pre-existing developed areas decreases the socioeconomic segregation allowing society to function more equitably, generating a tax base for housing, educational and employment programs.
- Encouraging community and stakeholder collaboration in development decisions.
- Shared municipal services that can provide services to more than one municipalities or communities and thereby reducing costs and taxes. These municipal services include the provision of potable water, wastewater treatment, garbage pickup and recycling, road maintenance, structural fire protection, and solid waste disposal.

Kunstler (1998) explains some basic principles for creating sustainable communities:

Design and Scale: The basic planning unit of any community, small or large, is still the neighbourhood. Even the largest cities have to be planned on a human scale, which includes social interaction, ease of mobility, access to goods and services including transit, and a sense of community. A neighbourhood can be defined as a 5-minute walking distance from a well-defined edge to a well-defined centre. This amounts roughly to an area of 1.3 km^2.

Connectivity: Neighbourhoods should be connected to other neighbourhoods and parts of neighbourhoods connected to other parts through corridors and districts. Corridors include streams, ravines, parks, natural areas, streets, railroads and footpaths. Districts are made up of several streets with a specific theme or activity in common, such as entertainment or business. It could be a neighbourhood dedicated to entertainment (theatres, cinemas, concert venues) in which housing, shops and offices are also integrated.

Diversity: Similar to natural ecosystems, social systems depend on diversity to ensure their overall health. In neighbourhoods, a mix of uses and activities is essential to the area's economic and social well-being. This includes a mix of stores and services near homes, with office space and public services (schools, library, post office) nearby. There is diversity of housing types as well, with apartments, single-family homes, duplexes and townhouses within a neighbourhood.

Attractiveness: Buildings should respect the aesthetics of public space, namely the street and its shade trees. Architectural design should match or improve the look of the area, and architectural guidelines could suggest appropriate exterior design features. Heritage buildings should be preserved wherever possible.

Networks: Streets are designed on a grid pattern to provide the greatest number of travel routes from one part of the neighbourhood to the other. This pattern improves overall traffic flow and reduces stresses on main streets. The grid should be modified by parks, squares, diagonals, T-intersections, roundabouts and other innovative designs. Cul-de-sacs and other roads disconnected from the whole are strongly discouraged. Preference should be given to civic buildings (town halls, schools, libraries) on prime building sites with high visibility and good access.

3.4 Restricting Urban Growth and Sprawl

Bengston et al. (2004) divided urban growth control policies into three main categories: *public acquisition*, *regulation*, and *incentives*. Public acquisition involves the purchase and ownership of lands by a public entity that subsequently prohibits development of the land. Areas targeted for purchase are often environmentally sensitive or have historical or recreational importance. Regulation includes the enforcement of local zoning laws, state management acts, and planning ordinances. On a small scale, regulation can mean specification of minimum building lot sizes.

On a large scale, it can include designating growth rate controls and placing moratoria on future development. Incentives and disincentives are also used for regulating urban growth. Developers might obtain benefits from creating projects that meet government planning goals. Conversely, developers might pay heavily for developments that have greater negative impacts on the community.[3] Incentives could include housing density bonuses for providing open space or tax credits for rehabilitating historic buildings. Disincentives could include development impact fees that are charged to builders in addition to normal permit fees (Bengston et al. 2004).

Several theoretical models to practical attempts and their evaluations can be seen towards the control of urban growth and sprawl (e.g., Easley 1992; Sasaki 1998; Brueckner and Lai 1996; Cooley and La Civita 1982; Engle et al. 1992; Hannah et al. 1993; Helsley and Strange 1995; Sakashita 1995; Staley et al. 1999; Ding et al. 1999; Fodor 1999; Couch and Karecha 2006; Lai and Yang 2002). The main question is how effective are the varied tools that have been developed, and implemented in many places towards controlling the urban sprawl? Alterman (1997), who compared the lessons learned in six nations, came to the conclusion that the effectiveness of most tools was moderate at best, since they cannot provide a solution to the problems facing agriculture vis-à-vis development pressures and decreasing profits. The effectiveness of growth management acts in reducing urban sprawl has been mixed (Anthony 2004; Kline 2000; Nelson 2000). Some researchers have observed that growth management plans might increase potential environmental damage in attempts to limit urban expansion (Audirac et al. 1990; Burby et al. 2001). However, interestingly nonetheless, even those who vigorously criticise restraining policy do support growth control implementation in order to prevent urban sprawl and reduce the conversion of open space and farmland (Frenkel 2004).

Frenkel (2004) has reviewed several existing tools for the control of urban growth and sprawl. Among the physical tools that were developed in order to restrain the conversion of farmland, one effective device is *exclusive agricultural zoning* (Coughlin 1991), which Oregon and Hawaii apply through state-wide control. An agricultural zoning is a designated portion of the municipality where agricultural uses are permitted and non-farm land-uses are either prohibited or allowed subject to limitations or conditions imposed to protect the business of agriculture. Alterman (1997) found that the most common tool across the United States was the *nonexclusive zone* which is much more flexible than agricultural zoning. Zoning may also include physical restrictions in dictating the land-use pattern in order to protect

---

[3] One popular approach to assist in smart growth in democratic countries is for law-makers to require prospective developers to prepare environmental impact assessments of their plans as a condition for state and/or local governments to give them permission to build their buildings. These reports often indicate how significant impacts generated by the development will be mitigated—the cost of which is usually paid by the developer. However, these assessments are often controversial. Conservationists and proponents are often sceptical about such impact reports, even when they are prepared by independent agencies and subsequently approved by the decision makers rather than the promoters. Conversely, developers will sometimes strongly resist being required to implement the mitigation measures required by the local government as they may be quite costly.

agricultural land (Hadly 2000). The idea of *green belt* (Longley et al. 1992), adopted mostly in Britain, though also found in other countries, is another popular means of setting limits to city expansion. A green belt is a land-use policy in which an area of agricultural (or other vegetation) land around an urban area is maintained to restrict the outgrowth of the city.

Annexation is another means of protecting non-urban periphery, by preventing open-space annexation to the city (Alterman 1997). Annexation incorporates a land area into an existing city or municipality, with a resulting change in the boundaries of the annexing jurisdiction. *Transfer of development rights* is another well-targeted tool to control the urban growth. It is based on the separation of the development rights of land from its ownership, thus enabling the transfer of these rights from the place where development is undesirable to other—more preferable and suitable places (Juergensmeyer 1984–1985; Anderson 1999; English and Hoffman 2001). *Infrastructure concurrency requirements* are recognised mostly in the progressive legislation adopted by Florida; it enables the local authority to determine the location and the timing of development in accordance with the timing of infrastructure development (DeGrove and Deborah 1992).

Some of the tools gathered from *adequate public facilities ordinances* basically support the justification of transferring the financial burden of public services to the new initiatives. According to Pendall (1999), this type of ordinance constitutes one of the most effective tools in restraining urban sprawl. Other forms of intervention have developed from this concept, among which are the following: *conservation easements* voluntarily transferring development rights to unprofitable organisation or funds in exchange for gaining tax rebates and easements; *purchase of development rights* (PDR), another kind of conservation easement. In this latter case, landowners sell their rights to develop land to the municipality for a limit number of years (Hadly 2000). A similar means that has recently been proposed is the *mitigation ordinance*. The initiator is forced to preserve a certain amount of its land in its natural condition against any amount of land to be converted for development (English and Hoffman 2001). More direct is *land purchase* by the government in order to preserve open space and farmland; this has been done, particularly in Oregon and Maryland, but in New Jersey, Florida, Pennsylvania, Arizona, Colorado and California as well (Hollis and Fulton 2002). Limited public funds are the major obstacle to a large operation of this tool.

Another important tool regarded as a means of targeting dispersed development and implemented in the United States under the growth-management framework is the *urban growth boundary* (UGB), which is widespread at the city level (e.g., Anderson 1999; Burby et al. 2001) and has been adopted widely as a planning instrument (discussed by Schiffman 1999) within a metropolitan or regional physical plan; for example, in Oregon and Minneapolis in the US, Melbourne in Australia, Santiago in Chile, and also in Israel amongst others (Nelson and Moore 1993; Asif and Shachar 1999; MSP 2003; Pendall et al. 2002; Wassmer 2002). UGB is a spatial boundary that controls urban expansion so that the land inside the boundary supports urban development and areas determined to remain non-urban outside the boundary. Empirical evidence on the impact of UGBs on development, however, is mixed. Some studies suggest that UGBs have effectively slowed down sprawling development (e.g., Kline 2005; Kline and Alig 1999; Nelson et al. 1997; Patterson 1999; Cho et al. 2006; Brueckner 2000, 2001; Bento et al. 2006), while others found that UGBs have no impact on growth (e.g., Staley et al. 1999; Jun 2004; Cho and Yen 2007). Detailed discussion on UGB can be found in Bhatta (2009a).

Infill growth is generally regarded as another remedy to urban sprawl. Obviously, the development of the inner vacant areas might prevent some peripheral development; but it is likely that the need and opportunities for peripheral development obviate the need for removing the obstacles to the development (Harvey and Clark 1965), especially in developing countries. Perhaps, the infill is likely to occur before the peripheral development. In case of high dense cities, majority of its residents generally seek some open space opposing infill. In such cases infill may cause serious challenge to human sustainability by increasing congestion and pollution. Furthermore, 100% saturation of built-up is neither possible nor it is desired for the health of the city.

Angel et al. (2005) argued that 'the merits of restricting urban expansion and encouraging infill and intensification of existing urban areas—even in the cities in industrialised countries—are by no means clear, nor is it self-evident that these are desired by the majority of urban residents. Even in the best of circumstances, compact city policies may have a marginal effect on the overall level of urban land consumption.' The case for densification and intensification in the cities of developing countries—where densities are, on average, three times higher than densities in industrialised and developed country cities—is even less clear.

However, whatever the criticisms one may find in the existing literature toward the urban growth and sprawl control policies, whether the impacts of these policies are marginal or widespread, most of the researchers shout for implementing these policies. To make these control policies more stringent, more than one policy can also be implemented simultaneously; for example, UGB in addition to *land purchase* by the government, or UGB with PDR. This may provide a new direction towards the control of urban sprawl.

Some investigations that have focused on different aspects of growth policy include:

- Studies of policy to increase open space, protect wildlife habitat, and limit environmental impact (Kline 2006; Radeloff et al. 2005; Dwyer and Childs 2004; Howell-Moroney 2004; Robbins and Birkenholtz 2003; Johnson 2001b; Bernstein 1994; Lewis 1990).
- Studies of how rural areas are confronting urban sprawl (Mattson 2002; Theobald 2005; Weiler and Theobald 2003).
- Investigations of agricultural land preservation efforts (Kashian and Skidmore 2002; Nelson 1992; Nelson 1990).

- Research to document the effects of growth on transportation. This includes modifying transportation plans to focus or channel growth (Buliung and Kanaroglou 2006; Kuby et al. 2004; Cervero 2001; Willson 1995).
- Research on alternative economic development through greening (Gatrell and Jensen 2002).

Chapter 4
Remote Sensing, GIS, and Urban Analysis

4.1 Introduction

Understanding urban patterns, dynamic processes, and their relationships is a primary objective in the urban research agenda with a wide consensus among scientists, resource managers, and planners; because future development and management of urban areas requires detailed information about ongoing processes and patterns. Central questions to be addressed are on how cities are spatially organised, where and when developments happen, why and how urban processes resulted in specific spatial pattern, and ultimately what may be the consequences of such pattern and/or process. Answers to these questions will definitely help us to prepare for the future in an equitable and sustainable manner; in specific, how the future planning should be done, whether to be more restrictive, how to overcome the ill effects of urban growth and sprawl, what policies should be appropriate in balancing the various competing goals of sustainability, and so on.

Remote sensing, although challenged by the spatial and spectral heterogeneity of urban environments (Jensen and Cowen 1999; Herold et al. 2004) seems to be an appropriate source of urban data to support such studies (Donnay et al. 2001). Detailed spatial and temporal information of urban morphology, infrastructure, land-cover/land-use patterns, population distributions, and drivers behind urban dynamics are essential to be observed and understood. Urban remote sensing has attempted to provide such information.

This chapter is aimed to discuss the confluence between remote sensing imagery, geographic information system (GIS) techniques, and urban analysis. It also sheds light on some issues associated with remote sensing and GIS for the analysis of urban growth.

4.2 Remote Sensing

In the world of geospatial science, *remote sensing* means observing the earth with sensors from high above its surface. They are like simple cameras except that they use not only visible light but also other bands of the electromagnetic spectrum such as infrared, microwave, and ultraviolet. Because they are so high up, these sensors can make images of a very large area, sometimes a whole province. Nowadays, remote sensing, also known as *earth observation*, is mainly done from space by using sensors mounted on satellites.

According to Jensen (2006)

> Remote Sensing is the non contact recording of information from the ultraviolet, visible, infrared, and microwave regions of the electromagnetic spectrum by means of instruments such as cameras, scanners, lasers, linear arrays, and/or area arrays located on platforms such as aircraft or spacecraft, and the analysis of acquired information by means of visual and digital image processing.

Remote sensing is a tool or technique that uses sophisticated sensors to measure the amount of electromagnetic energy exiting an object or geographic area from a distance and then extracting valuable information from the data using mathematically and statistically based algorithms. It functions in harmony with other spatial data-collection techniques or tools of the mapping sciences, including cartography and GIS. The synergism of combining scientific knowledge with real-world analyst experience allows the interpreter to develop heuristic rules of thumb to extract valuable information from the imagery (Bhatta 2008).

4.3 Urban Remote Sensing

It is irrefutable that *earth observation* is a modern science, which studies the earth's changing environment, through remote sensing data such as satellite imagery and aerial photographs. A report published by NASA highlighted the fact that the advances in satellite-based land surface mapping are contributing to the creation of considerably more detailed urban maps, offering planners a much deeper understanding of the dynamics of urban growth and urban sprawl, as well as associated matters relating to territorial management (NASA 2001).

Compared to other applications, remote sensing of urban areas, especially with space-borne sensors, is rather a new topic for the remote sensing community and geographers (Maktav et al. 2005). However, the interest and reliance on using remote sensing data in urban applications has shown a quantum increase. There are many reasons of strong reliance on remote sensing data in urban applications, such as, quick data acquisition over a large area, possibility of getting temporal datasets, advantages of digital processing and analysis, integration with GIS/GNSS, and many more (Bhatta 2008). Initially, sensors onboard aerial platforms dominated these applications but, currently, satellite-based sensors are competing. This development is the result of technical improvements that now allow satellite remote sensing systems to acquire images of very high spatial resolution.

Aerial photography has very long archived data records, while satellite remote sensing for earth observation started in 1972 with the first launch of Landsat satellite. Since 1972, numerous technical improvements have led to the second generation of earth observation satellites, such as advanced Landsat satellites (TM and ETM+), SPOT, and Indian Remote Sensing (IRS) LISS sensors. From 1999 one can distinguish a third generation of earth observation satellites with very high geometric resolution (IKONOS-2, QuickBird-2, OrbView-2, Geoeye-1, Cartosat, etc.). This has led to more and more urban applications using remote sensing data since the requirements regarding the desired level of detail can be fulfilled either by aerial or satellite-based sensor systems. Nowadays, not only these two platforms are complementary, rather satellite sensors are increasingly dominating many application domains including urban analysis.

In addition to the technical improvements of sensor systems, image analysis system also had to be improved. Continuous improvement and lowered price of computer hardware have made it available even to individuals. New strategies also needed to be developed for effective extraction of information. For example, the classical per-pixel classification is no longer the best methodology applied if one uses third generation high spatial resolution images. Therefore, object-orientated approaches or sub-pixel analysis, for example, proved to be better in many instances.

As a result of threefold development in software, hardware, and remote sensor technologies, there is an increasing demand among cities in developed countries for using remote sensing and GIS to establish an *urban information system*. Such a system needs to integrate ground measurements, traditional data (e.g., analogue maps and reports) and digital data (e.g. remote sensing images, digital maps, and attribute databases). This integration of various data sources allows better analysis of strategies that could result in better and timely planning and management of urban areas. However, this integrated approach to urban planning is not being applied effectively in developing countries due to the lack of data, and interaction and communication among planners, citizens and politicians (Maktav et al. 2005).

In terms of analysing urban growth, Batty and Howes (2001) believe that remote sensing technology, especially considering the recent improvements as mentioned above, can provide a unique perspective on growth and land-use change processes. Datasets obtained through remote sensing are consistent over great areas and over time, and provide information at a great variety of geographic scales. The information derived from remote sensing can help to describe and model the urban environment, leading to an improved understanding that benefits applied urban planning and management (Banister et al. 1997; Longley and Mesev 2000; Longley et al. 2001).

Remote sensing data are capable of detecting and measuring a variety of elements relating to the morphology of cities, such as the amount, shape, density, textural form, and spread of built-up areas (Webster 1995; Mesev et al. 1995). Remote sensing data are especially important in areas of rapid land-use changes where the updating of information is tedious and time-consuming via traditional surveying and mapping approaches. The monitoring of urban development is mainly to determine the type, amount, and location of land conversion. There are many studies on the use of remote sensing to monitor land-use changes and urban development

(Howarth 1986; Fung and LeDrew 1987; Martin 1989; Eastman and Fulk 1993; Jensen et al. 1993, 1995; Li and Yeh 1998). Remote sensing is very effective for illustrating the interactions between people and the urban environments in which they live (Gatrell and Jensen 2008). Space-borne satellite data are especially useful for developing countries due to the cost and time associated with traditional survey methods (Dong et al. 1997), and these techniques have become viable alternatives to conventional survey and ground-based urban mapping methods (Jensen et al. 2004a).

Several studies have demonstrated the applicability of remote sensing for supporting decision-making activities in a wide range of urban applications (Gatrell and Jensen 2008; Jensen and Cowen 1999; Zeilhofer and Topanotti 2008). In the areas of urban planning, many researches have been conducted using remote sensing imageries, particularly in urban change analysis and the modelling of growth (Bahr 2004; Hardin et al. 2007; Hathout 2002; Herold et al. 2003a; Jat et al. 2008; Jensen and Im 2007; Liu and Lathrop 2002; Maktav and Erbek 2005; Ridd and Liu 1998; Yang 2002; Yuan 2008), land-use/land-cover evaluation (Alphan 2003; Lopez et al. 2001; Xiao et al. 2006; Yang and Lo 2002; Yuan et al. 2005), and urban heat-island research (Kato and Yamaguchi 2005; Weng 2001). In particular, remote sensing based multi-temporal land-use/land-cover change data provide information that can be used for assessing the structural variation of land-use/land-cover patterns (Liu et al. 2003), which can be applied for avoiding irreversible and cumulative effects of urban growth (Yuan 2008), and are important to optimize the allocation of urban services (Barnsley and Barr 1996). Land-use/land-cover data derived from remote sensing data are also useful for devising sustainable urban growth and environmental planning strategies (Alphan 2003; Jensen and Im 2007).

4.4 Consideration of Resolutions in Urban Applications

Resolution (or resolving power) is defined as a measure of the ability of a remote sensing system or sensor to distinguish between signals that are spatially near or spectrally similar. Data collection system has four major resolutions associated with it. The major characteristics of an imaging remote sensing instrument are described in terms of its *spatial*, *spectral*, *radiometric*, and *temporal* resolutions.

If the composition of urban landscapes is considered, one can see that there are many small objects composed of many different materials in a spatial arrangement that does not give many homogeneous pixels in traditional earth observation images. Urban land-cover type changes frequently in a small transact. This determines that, for urban remote sensing applications, one has to consider the spatial resolution (to separate objects spatially), the spectral and radiometric resolution (to distinguish objects thematically) and the temporal resolution (to get the information on changes in the landscape through the time), according to the task to be fulfilled. Jensen (2006) or Bhatta (2008) may be referred for a detailed discussion on these resolutions.

4.4.1 Spatial Resolution

The spatial resolution or the ground resolution cell size of one pixel as the finite image element is the most important characterisation for a remote sensing image (Bhuyan et al. 2007). The detail discernible in an image is dependent on the spatial resolution of the sensor and refers to the size of the smallest possible feature that can be detected. Spatial resolution of passive sensors depends primarily on their *instantaneous field of view* (IFOV). The IFOV is the angular cone of visibility of the sensor which determines the area on the earth's surface that is 'seen' from a given altitude at one particular moment in time. The IFOV may also be defined as the area on the ground that is viewed by a single instrument from a given altitude at any given instant of time (Jensen 2006; Bhatta 2008).

The information within an IFOV is presented by a picture element in the image plane usually referred to as pixel. For a homogeneous feature to be detected, its size generally has to be equal to or larger than the resolution cell. If the feature is smaller than this, it may not be detectable as the average brightness of all features within that resolution cell will be recorded. However, smaller features may sometimes be detectable if their reflectance dominates within a particular resolution cell allowing sub-pixel detection (Yue et al. 2006; Xian and Crane 2005; Brown et al. 2000; Phinn et al. 2002).

The use of satellite imagery to map urban landscapes has met with varying degrees of success (Quattrochi 1983; Toll 1984; Duggin et al. 1986; Haack et al. 1987; Sadowski et al. 1987; Bhuyan et al. 2007). Buchan and Hubbard (1986) reported that even the 20 m multispectral resolution of the SPOT image is insufficient for mapping the heterogeneity of some inner city areas. In a low spatial resolution image, larger ground area makes a *mixed pixel* instead of homogeneous pixel. A mixed pixel is a pixel whose digital number (DN) represents the average energy reflected or emitted by several types of surface present within the area that it represents on the ground; sometimes called a *mixel*. This problem increases in the urban areas as because within a small transect the heterogeneity of land-cover is very high comparing with rural or other areas. A recent study (Bhatta 2007) shows that LISS–IV image of 5.8 m resolution also suffers from 15 to 20% overall confusion towards urban landscape classification due to mixed pixel and mixed class.

Important to mention that very high spatial resolution is also not preferred. Although higher spatial resolution provides better interpretability by a human observer; but a very high resolution leads to a high object diversity which may end up in problems when an automated classification algorithm is applied to the data.

The highest level of information needs in urban applications is individual blocks of buildings and requires the largest scales (1:1000–1:5000) since individual houses, roads, etc., are necessary to be detected in detail. The medium level would focus on a whole city and requires medium scales (1:10000–1:25000). The lowest level of detail focuses on regions, agglomerations and their surrounding areas and does not need a detailed differentiation within the city, therefore, requiring only small scales (1:50000–1:100000). For urban monitoring, in terms of urban growth detection, a scale of about 1:25000–1:50000 should be adequate (Sabins 1996). This application

Table 4.1 Application scale for various remote sensing images

Pixel size in m	Definition	Platform/Sensor[a]	Application scale
0.1–0.5	Extremely high res.	Airborne scanner, aerial photos, GeoEye-1 (pan), WorldView-1 (pan), WorldView-2 (ms)	1:500–1:5,000
>0.5–1	Very high res.	IKONOS (pan), QuickBird (pan), OrbView (pan)	1:5,000–1:10,000
>1–4	High res.	IKONOS (ms), QuickBird (ms), OrbView (ms), GeoEye-1 (ms), IRS (pan)	1:10,000–1:15,000
>4–12	Medium res.	IRS (pan), IRS (LISS-IV ms), SPOT (pan)	1:15,000–1:25,000
>12–50	Low res.	ASTER, IRS (ms), Landsat-TM/ETM+ (pan, ms), SPOT (ms)	1:2,500–1:100,000
>50–250	Very low res.	Landsat MSS	1:100,000–1:500,000
>250	Extremely low res.	NOAA	>1:500,000

[a] pan: panchromatic; ms: multispectral.

scale can be achieved from sensors which provide a spatial resolution of 4 m through 50 m (highlighted with grey in Table 4.1). However, in several instances Landsat-MSS (of 79 × 57 m spatial resolution) also considered in many urban applications owing to lack of higher resolution satellite imagery for the past.

> The Landsat sensor series does have a long lasting history of image acquisition. Actually it is the only series of remote sensing sensors acquiring imagery of a scale which is useful for urban monitoring. Since the Landsat series has been launched in 1972 with the first Multispectral Scanner (MSS), imagery has been recorded permanently and has been stored in data archives. Due to their more than 35 years period of operation these data are of very high value in terms of long term change detection monitoring. Most of the data are available for a low charge and handling fee, and even free for download through the internet. The websites for free Landsat image download are:
> http://glcf.umiacs.umd.edu/data/
> http://glovis.usgs.gov
> http://earthexplorer.usgs.gov

4.4.2 Spectral and Radiometric Resolutions

Not only the spatial resolution, but spectral and radiometric resolutions are also vitally important in urban applications. Many urban features have similar spectral characteristics; which means the objects give similar percent reflectance. This makes the object identification process more complicated.

Spectral resolution refers to the number and dimension of specific wavelength intervals in the electromagnetic spectrum to which a remote sensing instrument is sensitive. Different classes of features and details in an image can often be distinguished by comparing their responses over distinct wavelength ranges (Jensen 2006; Bhatta 2008). High spectral resolution is achieved by narrow band widths which, collectively, are likely to provide a more accurate spectral signature for discrete objects than broad bandwidths. Spectral resolution not only relates with the dimension of the bandwidth but also the number of bands. Individual bands and their widths will determine the degree to which individual targets can be discriminated on a multispectral image. The use of multispectral imagery can lead to a higher degree of discriminating power than any single band on its own. The ideal solution would be a hyperspectral scanner with a large number of bands each with a small bandwidth of 10 nm. But these remote sensing systems are limited and mainly in experimental stage. Furthermore, these sensors are recent advancement; therefore, for the analysis of past, hyperspectral images are not available.

It is important to mention that a lower spectral resolution does not create a mixed pixel; rather it creates *mixed class*, i.e., pixels representing different objects will belong to a single class. Class is a group of pixels relating to a narrow or broad category of objects over the earth surface.

The radiometric resolution is defined as the sensitivity of a remote sensing detector to differences in signal strength as it records the radiation flux reflected or emitted from the terrain (Jensen 2006; Bhatta 2008). It defines the number of just discriminable signal levels; consequently, it can have a significant impact on our ability to measure the properties of landscape objects. The radiometric resolution of an imaging system describes its ability to discriminate very slight differences in energy. The finer the radiometric resolution of a sensor, the more sensitive it is to detect small differences in reflected or emitted energy. Lower radiometric resolution is also responsible for mixed class instead of mixed pixel.

4.4.3 Temporal Resolution

Temporal resolution refers to how frequently the sensor can capture the images for a specific ground area (Jensen 2006; Bhatta 2008). It is another important factor especially when studying urban areas. An ideal solution would be a sensor system recording permanent the entire earth and delivering this data in real time. But due to technical restrictions such a system can not be realised. However, in many urban applications, especially for analysing the urban growth, high temporal resolutions are not required. For the analysis of urban growth, 5 yearly or 10 yearly intervals would be adequate. However, for some of the instances 15 yearly intervals are also accepted owing to lack of data. Most of the researchers prefer 10 yearly intervals to match with the census years that enable them to correlate the urban landscape with socioeconomic variables.

4.5 Geographic Information System

GIS is an information system designed to work with data referenced by spatial/geographical coordinates. In other words, GIS is both a database system with specific capabilities for spatially referenced data as well as a set of operations for working with the data. It may also be considered as a higher-order map or an intelligent map on which computer analysis can be performed.

GIS has many alternative definitions (Bhatta 2008); however, from the perspective of urban growth a good definition of GIS may be given as:

> an information system that is used to input, store, retrieve, manipulate, analyse and output geographically referenced data or geospatial data, in order to support decision making for planning and management of land-use, natural resources, environment, transportation, urban facilities, and other administrative records.

A GIS can integrate a variety of data, but most data layers in a GIS are considered 'geospatial' (or 'spatial') data because they are associated with specific locations on the earth's surface (georeferenced) and linked to additional information (attributes) about that location. GIS is a system that can not only integrate a variety of data, rather it is an integration of several subsystems (Fig. 4.1). Bhatta (2008) describes GIS as a set of interrelated subsystems and as a knowledge hub.

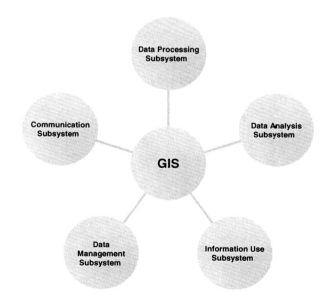

Fig. 4.1. GIS is a set of interrelated subsystems

4.5 Geographic Information System

A GIS has three different views: (1) *database view*—fundamentally, a GIS is based on a structured database that describes the world in geographic terms; (2) *map view*—a GIS is a set of intelligent maps and other views that show features and feature relationships on the earth's surface; and (3) *model view*—a GIS is a set of information transformation tools that derive new geographic datasets from existing datasets (in other words, by combining data and applying some analytic rules, we can create a model that helps answer the question we have posed).

4.5.1 GIS in Urban Analysis

The advantages of using GIS for urban analysis are many. GIS, as a database management system, offers forward data mapping functions for displaying geographical information, and backward data retrieval functions for 'querying' maps (Levine and Landis 1989). These 'front end' and 'back end' operations allow analysts and planners to better manage, display and communicate information (Miller 1999). The power of these functions, furthermore, is augmented by techniques for interactive data modelling (e.g., cartographic analysis, data conversion routines), which can enhance urban transportation analysis (e.g., Wang and Cheng 2001; Stanilov 2003) and land-use analysis (e.g., Landis and Zhang 1998). In some of the instances, use of GIS techniques for urban analysis may even enable a researcher to break 'free' from the 'tyranny of zones' (Spiekermann and Wegener 2000). However, despite its considerable advantages, GIS in and by itself does not 'free' the analyst from the necessity of coping with the nuances, complexities and subtle relationships inherent in the spatial data commonly used in urban research (Páez and Scott 2004).

Regarding the complexities of modelling spatial data, a paper by Fotheringham (2000) poses the question of whether the adoption of GIS has represented a step forwards or a step backwards for spatial modelling. In answering the question, he asserts that, to date, most GIS-based modelling represents a step backwards—although commercial GIS packages now incorporate spatial models, in addition to cartographic analysis techniques, these models tend to be outdated and far away from the research frontier. In case of urban analysis, the adoption of GIS has given the strong impression that the discipline embraces space in modelling. It has been noted, however, that the statistical models underlying conventional urban analysis remain, for the most part, aspatial (Landis and Zhang 2000), and thus ignore important issues such as the failure of most conventional statistics to adequately summarise locational information (the *sufficiency criterion*; Griffith 1988), the lack of independence and inherent stickiness of spatial data (e.g., *spatial association*), differential effects in spatial processes (e.g., *heterogeneity*), and the implications of shape and representation in spatial analysis (Páez and Scott 2004). It is thus surprising, given the seriousness of these issues (Anselin and Griffith 1988), that compared to the swift adoption of GIS, urban analysts have been slower to embrace technical developments in spatial analysis. This is, however, starting to change.

4.6 Urban Analysis

Urban analysis has been defined as the use of multidisciplinary knowledge and skills with the objective of solving urban problems (Pacione 1990). The term 'urban systems analysis' can be applied to the study of an individual city, conceptualised as a collection of various interrelated components (Páez and Scott 2004). Typically, these components include an activity sub-system that determines a city's land-use configuration, transportation sub-system and the interactions between these components (Black 1981; De la Barra 1989; Kanaroglou and Scott 2002). More recently, urban analysts have paid increasing attention to the urban environment and the effect of land-use and transportation activities on it. A comprehensive list of urban sub-systems and their interactions can be found in Moeckel et al. (2003).

Processes of interest in urban analysis include different types of construction (residential, industrial, transportation infrastructure), economic and demographic changes, mobility (travel, residential choice and freight) and environment-related processes (such as energy consumption, emission of pollutants and noise). Methodological issues include the definition of units of analysis, which could be aggregated (e.g., traffic analysis zones) or disaggregated (e.g., individuals or households) (Páez and Scott 2004). A characteristic of most urban processes is the fact that they are spatial in nature and therefore space-dependent.

The analysis of urban systems has a long history using the computer for modelling and simulation. As recounted by Harris (1985), urban simulation, since its inception in the 1950s, was enabled and stimulated by developments in digital computing. Furthermore, the influence of the computer was both direct and indirect through the role that digital computing played in making possible the present scope of statistical and other methods used in urban geographic research (e.g., mathematical programming). The history of urban analysis is thus intertwined with that of the computer, on the one hand, and the development of statistical and mathematical methods on the other. A recent expression of the long tradition in urban analysis of rapidly embracing technological developments can be observed in the adoption of remote sensing data and GIS (Páez and Scott 2004).

The analysis of urban system may include a wide spectrum of research—e.g., health research (e.g., Robinson 2000; Jerrett et al. 2003), urban crime (e.g., Craglia et al. 2000; Craglia et al. 2001; Ceccato et al. 2002), political-historical processes (e.g., Flint 2002), and merging urban ecology and socioeconomics (McDonnell and Pickett 1990; Grove and Burch 1997; McDonnell et al. 1997; Zipperer et al. 1997; Gomiero et al. 1999; Zhang et al. 2006; Uy and Nakagoshi 2008); estimation of urban population (Zhang 2003); urban change analysis and the modelling of growth (Bahr 2004; Hardin et al. 2007; Hathout 2002; Herold et al. 2003a; Jat et al. 2008; Jensen and Im 2007; Liu and Lathrop 2002; Maktav and Erbek 2005; Ridd and Liu 1998; Yang 2002; Yuan 2008); land-use/land-cover evaluation (Alphan 2003; Lopez et al. 2001; Xiao et al. 2006; Yang and Lo 2002; Yuan et al. 2005); urban heat-island research (Kato and Yamaguchi 2005; Weng 2001); and many others. Gatrell and Jensen (2008) cite several references to describe how remote sensing data have been used in urban areas, and highlight some research areas where remote sensing may continue to aid urban geographical inquiry and some potential pitfalls. They conclude that remote sensing has a critical role to play in the analysis of the interactions that occur between people and urban environments that may help shape our understanding of humans and the environment in which they live.

Based on the foundations of urban remote sensing, a new collection of researchers have begun to ask novel questions concerning the socio-spatial implications of the observed interactions between the built and natural environments with respect to observed socioeconomics as well as related implications on public policy (Jensen and Cowen 1999; Gatrell and Jensen 2002; Heynen and Lindsey 2003; Jensen et al. 2004b; Malczewski and Poetz 2005; Mennis and Jordan 2005). Specifically, researches have evaluated how urban conditions and associated socio-demographics correlate at multiple scales. More recently, researchers have begun to link observed quality of life indicators with remotely sensed data to construct a better understanding of basic human–environment interactions.

However, the primary focus in this book is on the analysis of urban growth, and it excludes many important analytical methods and models that are practiced in the analysis of urban systems.

4.6.1 Analysis of Urban Growth

Urban growth is a spatially-conditioned process, and the outcome at one location is partially affected by the events of its neighbouring locations. Several spatial processes relevant to urban growth analysis can be identified in the existing literature of spatial data analysis. Determining the rate of urban growth and the spatial configuration from remote sensing data, is not only a prevalent approach, rather has a long history (Donnay et al. 2001; Maktav et al. 2005; Herold et al. 2005a; Batty 2000). GIS also has a wide spectrum of applications in urban analysis, because the discipline embraces space in analysis and modelling (Fotheringham and Wegener 2000; Okunuki 2001; Du 2001; Abed and Kaysi 2003; Liu and Zhu 2004; Barredo et al. 2004).

Morphological measurement of urban development is important for land-use planning and the study of urban development (Webster 1995). Urban planners and researchers are often concerned with the change in size, shape, and configuration of built-up area. The measurement of urban form can provide a more systematic analysis of the relationships between urban form and process. Urban economics, transport, and social structure are predicted in spatial terms, and thus, the effects of such a theory are often articulated through geometric notions involving the shape of urban land-use and the manner in which it spreads (Mesev et al. 1995).

Banister et al. (1997) found that there are significant relationships between energy use in transport and physical characteristics of a city, such as density, size, and amount of open space. Batty and Kim (1992) have shown an interesting approach in the measurement of self-similarity using fractal analysis. Fractal geometry can provide a more insight into density functions. It provides ways in which the form of development can be linked to its spread and extent (Mesev et al. 1995). Shape is also a very important element considering the optimal location of development. It can be used to find the best location for a particular land-use by the use of GIS techniques (Brookes 1997).

The spatial configuration and the dynamics of urban growth are important topics of analysis in the contemporary urban studies. The review of literature found that the urban growth has aroused wide social focus because in many instances this growth is uncontrolled and sprawled which may impede regional sustainable development. Rapid urban growth in the world is quite alarming, especially, in developing countries (Angel et al. 2005; Kumar et al. 2007). Therefore, the importance of conducting researches on urban growth has been strongly felt throughout the world; and related studies have come out consequently which mainly cover the pattern, process, causes, consequences and countermeasures.

However, this book deals mainly with the pattern and process of urban growth. Urban growth should be analysed both as a pattern of urban land-use—i.e., a spatial configuration of a metropolitan area in a specific time—and as a process, namely as the change in the spatial structure of cities over time. Urban growth as a pattern or a process is to be distinguished from the causes that bring such a pattern about, or from the consequences of such patterns (Galster et al. 2001). If the urban growth is considered as a pattern it is a static phenomenon and as a process it is a dynamic phenomenon. Analysis of urban growth, as a pattern or process, is an essentially performed operation by the city planners and administrators, proponents, and other stakeholders. However, stakeholders are generally interested in the future pattern of urban land-use rather than the past or present in view of their investment goals; but the city administrators/planners and proponents require analysing the pattern of urban growth for the past and present in order to prepare for the future (Bhatta et al. 2010a).

Analysis of urban growth, as a pattern and process, helps us to understand how an urban landscape is changing with time. This understanding includes (1) the rate of urban growth, (2) the spatial configuration of growth, (3) whether there is any discrepancy in the observed and expected growth, (4) whether there is any spatial or temporal disparity in growth, and (5) whether the growth is sprawling or not (Bhatta et al. 2010a). As mentioned earlier, the analysis of urban growth can be performed for the past, present, and the future as well. However, the case of past and present is different than the future; because the former is based on empirical evidences whereas the later is based on simulations. However, simulation is obviously dependent on the past and present evidences.

In the recent years, remote sensing data and GIS techniques are widely being used for mapping (to understand the pattern), monitoring (to understand the process), measuring (to analyse), and modelling (to simulate) the urban growth, land-use/land-cover, and sprawl. The physical expressions and patterns of urban growth and sprawl on landscapes can be detected, mapped, and analysed by using remote sensing data and GIS techniques (Pathan et al. 1989, 1991; Barnes et al. 2001; Angel et al. 2005; Kumar et al. 2007). The decision support systems within the GIS can evaluate remote sensing and other geospatial datasets by using multi-agent evaluation (Axtell and Epstein 1994; Parker et al. 2003) which can also predict the possibilities in the subsequent years using the current and historical data. In the last few decades, these techniques have successfully been implemented to detect, analyse, and model the urban growth dynamics.

4.6.2 Analysis of Urban Growth Using Remote Sensing Data

Remote sensing data are able to address both time and space considerations at a range of spatio-temporal scales. Two fundamental means of conceptualising scale are in terms of data/information frequency, and in terms of data/information range (Prenzel 2004). From the spatial perspective, frequency refers to pixel size (spatial resolution) while range refers to the area covered by the image. From the temporal perspective, frequency refers to temporal intervals at which images are obtained (or considered for analysis) and range refers to the time period spanned by the image data set (the time span under analysis). *Spatial frequency*, *spatial range*, *temporal frequency* and *temporal range* have a fundamental bearing on how remote sensing based analysis is performed in relation to urban change. Analytical and simulation models are certainly dependent on these spatio-temporal scales. The choice of model, in many instances, is influenced by these scale parameters. For example, a very high spatial resolution increases the heterogeneity in the image, and therefore may not be suitable for patch metrics analysis since the number of patches will be extremely high resulting in difficulties in understanding the landscape in general.

A useful paradigm for conceptualising urban change analysis is in terms of the degree-of-indirectness with which the measured radiation field and the wanted quantity are linked (Quenzel 1983). Within this framework, analysis models are arranged according to a particular 'model order'. As the order of the model increases, the link between the measured radiation field and the quantity of interest become more remote. The approach is more conceptual than pure physics, but it gives some insight into the urban remote sensing problem (Quenzel 1983). The framework enables one to conceptualise the level of accuracy achievable for the quantities derived from the remote sensing imageries, because error increases as the link becomes more complex (i.e., as the model order increases).

Examples (from Prenzel 2004) of a first-order analysis are the extraction of surface reflectance from raw radiance values, whereas a second-order analysis would involve the extraction of some type of measured surface characteristic, such as surface temperature or ozone concentration in the atmosphere. Third-order analysis derives several biophysical parameters, for example, land-cover, leaf area index, biomass, or leaf pigment content. Fourth-order analysis derives parameters—for example—fuzzy and thematic measures of land-use, measures of net primary productivity. The difficulty with accurately extracting high-order parameters such as land-use is that they are not closely linked to the physical properties of terrain that are measured by remote sensing process.

A second and complementary way of describing remote sensing analytical models is in terms of their 'determinism' (Schott 1997). Generally, these models can be thought of as occurring along a continuum between being deterministic and being empirical (i.e. statistical). The basic difference between these two concepts is that deterministic processes can be quantitatively described and predicted very accurately (i.e. close to 100%), whereas empirical processes can only be quantitatively described in terms of levels of confidence. It may be worth mentioning that a well developed deterministic model can be inverted to accurately predict biophysical variables from the original remote sensing data beyond the confines of the image from which the model was developed (Quenzel 1983). Determinism is the view that every event, including human cognition, behaviour, decision, and action, is determined. In contrast, empirical models cannot be used to accurately predict parameters beyond the image data from which they were derived. Perhaps, the main drawback of deterministic models is that they are typically more demanding in terms of data inputs, model establishment, and model validation (Jensen 2005). Whereas, empirical models are often easier to develop and apply since they are generally less demanding in terms of analysis, model establishment, and validation.

Empirical methods include: (i) *image subtraction*, (ii) *classification*, (iii) *neural network approaches*, (iv) *image index* (ratioing), amongst others. Empirical approaches have been applied in a variety of mapping contexts; for example, *forestry* (Verbyla and Richardson 1996), *urban–rural fringe* (Gao and Skillcorn 1998), *agriculture* (Salem et al. 1995), *land-cover* (Langford and Bell 1997), and *land-use* (King 1994). Deterministic-oriented analyses have been used to characterise: (i) *chlorophyll content* (Zarco-Tejada et al. 2001), (ii) *leaf area index* (Chen et al. 1997), (iii) *chlorophyll fluorescence* (Zarco-Tejada 2000), (iv) *nitrogen content* (Johnson 2001a), and (v) *net primary productivity* (Gower et al. 1999). Empirical and deterministic models can also be mixed that are neither strongly empirical nor deterministic. Mixed models are confluence of empirical and deterministic approaches. Some examples are: (i) *linear spectral unmixing* (e.g. estimating biophysical forest parameters; Peddle et al. 1999), (ii) *non-linear spectral unmixing* (e.g. mapping dense vegetation; Ray and Murray 1996), and (iii) *physically based classification* (e.g. land-cover mapping; Zarco-Tejada and Miller 1999). Mixed models generally use the *spectral mixture analysis* technique to extract change information.

Urban change or growth analysis, by definition, requires comparison of land-surface information through time but it does not necessarily require comparison through space beyond the confines of the image being analysed (Prenzel 2004). Remote sensing derived land-surface information required for planning, in several instances, is of high-order. However, in this sense, 'high-order' means a reduction in the degree to which results can be compared or 'inverted' to other images (Quenzel 1983). Although the ability to compare land surface change information across the space is primary concern for many planning activities, however, still there are many cases where high-order information is needed. The basic question that arises from this discussion is whether to use deterministic or empirical approaches in the context of urban growth analysis.

4.6 Urban Analysis

Traditionally, two very general change detection applications using remote sensing data are identified, namely: (1) biophysical monitoring for establishment and calibration of mass/energy models; and (2) context-specific land-surface monitoring (Miller and Chen 2001). The former mainly corresponds to deterministic model and the later is linked with empirical models. Urban growth analyses may not require transfer of information across space beyond the confines of the image being analysed. In such cases, temporal consideration may be more important than the understanding of space beyond the image data. Therefore, in many instances, empirical models, rather than purely deterministic models, may be more useful for providing site specific urban change information.

Chapter 5
Mapping and Monitoring Urban Growth

5.1 Introduction

Mapping of urban growth is different then the mapping of urban area. Urban area can be detected and mapped using a single temporal image, that is, image from a specific date. However, mapping of urban growth necessitates minimum of two temporal imageries; since it actually means the mapping of changes between two different dates. This process is often called *change detection*—the process of identifying differences in the state of an object or phenomenon by observing it at different dates (Singh 1989). Mapping of urban growth can lead to motoring of urban area and its growth through time.

In general, urban change detection involves the application of multi-temporal datasets to quantitatively (or visually) analyse the temporal effects of the phenomenon (Lu et al. 2004). Various techniques have been developed to improve change-detection accuracy, including image differencing, image ratioing, post-classification comparison, the masking method, principal component analysis, etc. Table 5.1 lists several methods of urban growth mapping, from which, some popular methods will be discussed in this chapter. It is worth mentioning that no single change detection approach can globally be recommended.

5.2 Image Overlay

Image overlay is one of the simplest methods of visually identifying the location and extent of change or urban growth (Jensen 2005). It is generally done by passing single bands of imagery from different dates through the red, green, and blue (RGB) colour guns[1] of a colour monitor for displaying the image (Banner and Lynham 1981). To identify the changes between images from two different dates, a single band (from old date—t_1) can be passed through both the blue and green

---

[1] More discussion on display of multispectral image on a colour monitor can be found in Bhatta (2008).

Table 5.1 A summary of urban change detection methods

Method	Reported usage
Change vector analysis	Chen et al. (2003a) and Johnson and Kasischke (1998)
Decision trees	Im and Jensen (2005) and Chan et al. (2001)
Econometric panel	Kaufmann and Seto (2001)
Image differencing	Todd (1977), Toll et al. (1980), Quarmby and Cushnie (1989), Fung (1990), Ridd and Liu (1998), Sunar (1998), Bruzzone and Prieto (2000), Masek et al. (2000), Maktav and Erbek (2005) and Liu et al. (2004)
Image ratioing	Nelson (1983), Maktav and Erbek (2005) and Liu et al. (2004)
Image regression	Ridd and Liu (1998) and Liu et al. (2004)
Kauth-Thomas image differencing (tasseled cap transformation)	Fung (1990), Ridd and Liu (1998), Kaufmann and Seto (2001) and Seto and Fragkias (2005)
Learning vector quantization	Chan et al. (2001)
Artificial neural networks	Dai and Khorram (1999), Chan et al. (2001), Liu and Lathrop (2002), Pijanowski et al. (2005) and Bruzzone (1999)
Principal components analysis	Fung and LeDrew (1987), Fung (1990, 1992), Yeh and Li (1997), Li and Yeh (1998), Sunar (1998), Liu et al. (2004) and Fung and LeDrew (1987)
Post-classification comparison	Howarth and Wickware (1981), Fung (1992), Jensen et al. (1995), Li and Yeh (1998), Sunar (1998), Ward et al. (2000), Chan et al. (2001), Madhavan et al. (2001), Yang (2002), Yang and Lo (2002), Chen et al. (2003a, b), Weber and Puissant (2003), Alberti et al. (2004), Mundia and Aniya (2005), Xiao et al. (2006) and Yu and Ng (2006)
Spectral-temporal classification	Schneider et al. (2005)
Vegetation index comparison	Maktav and Erbek (2005) and Townshend and Justice (1995)

colour guns, while the same spectral band (for the second or new date—t_2) is passed through the red colour gun. Land-cover change that is responsible for the change in reflectance in the selected bands will appear bright red or cyan,[2] while areas with little or no-change will appear in grey shades (Fig. 5.1). For detecting the changes among imageries of three dates, it is required to pass corresponding bands through all three RGB colour guns. Resulting colours/shades are then cyan, yellow, magenta, and grey; where, grey indicates little or no-change, and others indicate change in land-cover.

This is a simple approach and may be used for quick identification of changes in urban landscape, particularly when monitoring urban fringe areas where conversion

---

[2] Changes will appear in bright red if the pixel reflectance value increases due to change. In case of decrease in reflectance value, changes will appear in cyan colour instead of red.

5.2 Image Overlay

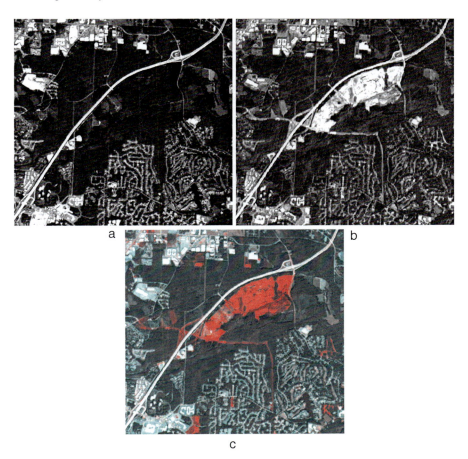

Fig. 5.1 (**a**) Image from date t_1 which have been passed through *blue* and *green* colour guns; (**b**) Image from data t_2 which have been passed through *red* colour gun; (**c**) Urban growth appears as *bright red* in the composite image

of vegetation to impervious surface produces a large change in spectral reflectance. Although image overlay is incapable to quantify the changes, it allows the analysts to visually interpret its extent and location. Subsequent quantitative procedures can then be adopted for further understanding. By noting the bands in which changes can be identified easily, image overlay is also a rapid way to select appropriate bands for inclusion in image differencing (Sect. 5.3). This method does not require specialised image processing software and is easy to understand by those unfamiliar with image processing (Hardin et al. 2007). Finally, this process is computationally least expensive.

The greatest limitation of this approach is that one can not quantify the change or identify the transition among different land-cover classes (i.e., what has changed to urban). This method of change detection requires high-accuracy image registration.[3] Further, this process does not create a change map; therefore, changes to be mapped by manual process. Mapping the change is also difficult because determining the line that differentiates the change and no-change often ambiguous. Finally, manual mapping of changes for a large area is time consuming and labour intensive.

5.3 Image Subtraction

Image subtraction (also called *image differencing*) uses software algorithm to identify and quantify the changes between two temporal images. Typically, two images which have been geometrically registered are used with the pixel values in one image being subtracted from the pixel values in the other. In such a *change image* (or *difference image*), areas where there has been little or no-change between the original images contain resultant brightness values around 0, while those areas where significant change has occurred contain values higher or lower than 0, e.g., brighter or darker depending on the 'direction' (positive or negative) of change in reflectance between the two images (Fig. 5.2a). This is one type of image transform that can be useful for mapping changes in urban development around cities and for identifying areas where deforestation is occurring (Bhatta 2008). The pixel value in the difference image (Dx_{ij}^k) can be calculated as,

$$Dx_{ij}^k = x_{ij}^k(t_2) - x_{ij}^k(t_1) \tag{5.1}$$

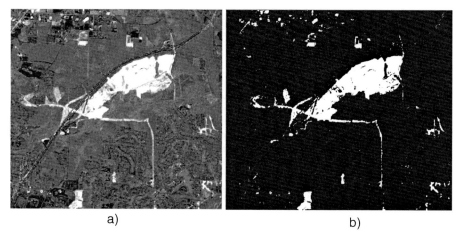

Fig. 5.2 (a) Difference image obtained from image subtraction (original images are shown in Fig. 5.1a and b); (b) Mask image after thresholding (urban growth appears as *white*)

---

[3] Image registration is fitting of the coordinate system of one image to that of a second image of the same geographic area.

5.3 Image Subtraction

where, x_{ij}^k = pixel value for band k, row i and column j in the image, t_1 = first date, t_2 = second date.

The difference image can also be used to create a binary mask of change vs. no-change (Fig. 5.2b). This mask $M(x)$ is created by defining a threshold value (T) such that (Singh 1989; Hardin et al. 2007):

$$M(x) = \begin{cases} 1, & \text{if } |D(x)| > T \\ 0, & \text{otherwise} \end{cases}; \text{ where } D(x) = \text{difference image}, x = \text{pixel value} \quad (5.2)$$

The choice of threshold is crucial. It is selected to separate relevant and real changes from those irrelevant or created by seasonal/diurnal effect or noise. If the threshold value is too small, the difference mask will overestimate the area of change; and a high threshold value would exclude many areas of change from the mask.

It is also possible to produce separate change masks for different types of change through the use of density slicing (Singh 1989; Pilon et al. 1988). Such change masks can provide information on positive change, negative change, and no-change. Some of the image processing software (e.g., ERDAS Imagine) automatically slices the change image based on the user-defined threshold values.

The analyst usually determines the threshold value empirically (Jensen 2005; Singh 1989). To set thresholds that make the change map meaningful, the analyst needs to understand both the target and the project goals (Schowengerdt 1997). Researchers have also proposed statistical methods of selecting the threshold (e.g., Ingram et al. 1981; Fung and LeDrew 1988; Singh 1989; Yuan et al. 1999; Morisette and Khorram 2000; Smits and Annoni 2000; Rosin 2002; Rosin and Ioannidis 2003; Coppin et al. 2004).

Despite of having these statistical aids, problems in interpreting the magnitude of change often make it impossible to define a global threshold value. For example, a reflectance difference of 30% in the urban periphery may indicate conversion from agriculture to urban residential land, while the same reflectance change in the CBD (central business district) may result from street repaving (Hardin et al. 2007).

Input images having pixels of vastly different magnitudes (intra image) can also create the same difference values for different type of changes in the subtraction process. This is a normal case in urban landscapes since they are highly heterogeneous in nature. Although reflectance magnitudes on the two images would indicate different change-class involvement, change classes that generate same pixel values in the difference image can not be differentiated. Coppin and Bauer (1994) suggest dividing the difference by the sum of the input values to mitigate this problem. However, this technique is also not adequate, in many instances, to identify different changes individually.

In addition to the aforementioned hurdles, the technique is sensitive to noise and temporal (seasonal/diurnal) illumination differences that make it a questionable choice for urban change detection. It is, in specific, a poor choice when high resolution imageries are in use where shadows and slight illumination change causes large change in radiance recorded by the sensor.

Finally, image subtraction uses single band of the multispectral imagery; therefore, full utilisation of spectral information is not possible. It is obvious that many earth surface features give similar reflectance in a specific band of electromagnetic spectrum; and the use of multispectral information would be the only choice to discriminate them.

The advantages of image subtraction method are (1) quick analysis using computer software, (2) quantification of change, and (3) automated change map generation.

5.4 Image Index (Ratioing)

Image index or *ratioing* is another method of change detection. Although image index has not been as intensively investigated as image differencing (Nelson 1982), it is an effective method of change detection. This method uses one temporal image to divide image of another date:

$$\text{Pixel value in ratio image } (Rx_{ij}^k) = \frac{x_{ij}^k(t_1)}{x_{ij}^k(t_2)} \tag{5.3}$$

The resulting per-pixel quotients constitute a ratio image with image pixels unchanged between image dates producing a value of 1.0 and areas of change giving higher or lower values depending on the change-classes involved (Fig. 5.3). For example, if we divide a near-infrared (NIR) band of old image by the NIR band of

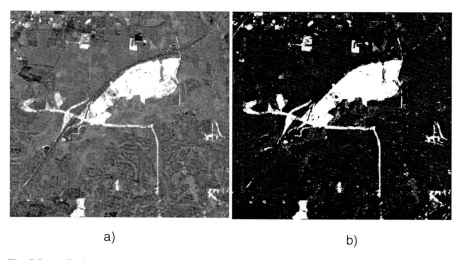

Fig. 5.3 (a) Ratio image (original images are shown in Fig. 5.1a and b); (b) Ratio image has been enhanced by changing the *brightness/contrast* to highlight the urban growth

5.4 Image Index (Ratioing)

new image, values near to 1.0 indicate no-change and values greater or lesser than 1.0 indicate changes. In this case values greater than 1.0 may indicate urban growth by destroying vegetation.

The critical element of the methodology is selecting appropriate threshold values in the lower and upper tails of the distribution representing change pixel values. The usual practice has been in selecting arbitrary threshold values and testing them empirically to determine if the change detection was performed accurately (Nelson 1982). However, because the output quotient distribution is non-Gaussian and bi-modal, it is impossible to set a reliable numeric threshold of meaningful change (Robinson 1979; Coppin and Bauer 1994). If the distributions are non-normal and functions of the standard deviations are used to delimit change from non change, the areas delimited on either side of the mode are not equal; therefore, the error rates on either side of the mode are not equal (Singh 1989). Nevertheless Robinson (1979) recommends that the further studies of the ratioing method under a variety of conditions would be useful.

Although, image index was first used in vegetation studies with Landsat Multispectral Scanner data (Rouse et al. 1974), it has successfully been used in an urban landscape analysis by Todd (1977) who has reported 91.4% accurate classification of changes. Image indexing has similar limitations and advantages as with the image subtraction (Sect. 5.3).

Image subtraction and index may be performed on multi-temporal images that have previously been transformed to enhance important change classes, rather than on original image bands. Some examples of these transformations are (Hardin et al. 2007):

- Kauth-Thomas or tasseled-cap transform (Ridd and Liu 1998; Rogan et al. 2002)
- Normalized differenced vegetation index (NDVI) (Cakir et al. 2006; Du et al. 2002; Masek et al. 2000; Song et al. 2001)
- Principal component analysis transform (Du et al. 2002; Millward et al. 2006)
- Miscellaneous indices and transformations, for example, Schott et al. (1988) found that an infrared to red ratio is useful in separating urban pixels from water and vegetation.

For example, Nelson (1982) has used ratio vegetation index to generate difference image (Dx):

$$Dx = \frac{\text{near infrared band}}{\text{red band}}(t_1) - \frac{\text{near infrared band}}{\text{red band}}(t_2) \qquad (5.4)$$

This difference image provides an avenue for deciding whether or not a vegetation canopy has been significantly altered (Nelson 1982). Angelici et al. (1977) also used the difference of ratio data and the thresholding technique to delineate changed areas.

A number of transformations can also be conducted in combination. The selection of transformation is often dependent upon the scene properties and project requirements. The appropriate method is the one that works to solve the problem; and determining the right method is a process of trial and error. This approach is more complex and beyond the understanding of general image analysts. If the tool is more complex than the purpose for which the tool is being used then its applicability comes under question that restricts widespread utilisation.

5.5 Spectral-Temporal Classification

Another method for change detection is *spectral-temporal classification* (Schneider et al. 2005). This method uses a composite image created by adding (i.e., 'stacking') the bands from multiple dates of imagery together to form a single image. If two four-band images are stacked, the output is a composite image of eight bands. This image is then classified using traditional image classification techniques to identify change classes.

Despite of being an effective method of change detection it has several limitations. The classified image results in many classes that are often difficult to interpret and not effective for cartographic representation. Signature collection procedure for supervised classification is also difficult. Image overlay technique may be used for the collection of signatures, but, image overlay technique itself suffers from identification of what has changed and changed to what. As a result, unsupervised approach is most preferred which requires image to be classified in large number of classes and to be identified accordingly. Since the composite image has many bands from different dates, there will be a high spectral confusion as well, resulting in difficulties in classification. Further, temporal images should have same spatial resolution and similar spectral characteristics that are often difficult to obtain; especially if the analysis is based on long temporal span.

Important to realise, the image to be classified has many bands, some of which may be redundant in information content (Estes et al. 1982). The problem of redundancy can be overcome by using a principal component transformation on the original data set. The first few components containing significant amounts of variance from the two dates can be used in the classification. Another problem is that the temporal and spectral features have equal status in the combined data set (Schowengerdt 1983). Thus spectral changes within one multispectral image cannot be easily separated from temporal changes between images in the classification. Swain (1978) developed a Bayesian (minimum risk) 'cascade' classifier to remove this coupling between the spectral and temporal dimensions. In this classifier, preliminary classification is done at time t_1, and, when data become available, from time t_2, transition probabilities and t_2, likelihood values are determined and an updated classification is made. However, this method has not attracted further attention, apparently because of the complexity of the algorithms and computational requirements (Singh 1989).

5.6 Image Regression

Change detection through *image regression* assumes that pixel values recorded for the same location on two different dates are linearly related. To detect the changes, a linear least-squares regression is performed using the pixel values of image at time t_1 as the independent variable and the corresponding pixel values at t_2 as the dependent variable. The change map is a map of regression residuals. It is created by subtracting the regression-predicted t_2 reflectance values from the actual t_2 pixel reflectance values. Pixels with large residuals (based on a threshold) are used to generate a change/no-change mask as explained in Sect. 5.3 (Singh 1989).

Mathematically, one can regress $x_{ij}^k(t_1)$ against of $x_{ij}^k(t_2)$ using a least-squares regression and can predict value of corresponding pixel at t_2. If $X_{ij}^k(t_2)$ is the predicted value obtained from the regression line, the difference image can be defined as follows:

$$Dx_{ij}^k = X_{ij}^k(t_2) - x_{ij}^k(t_2) \tag{5.5}$$

Image regression has some theoretical advantages as documented by Hardin et al. (2007); for example, (1) it accounts for differences in reflectance mean and variance between dates, (2) it also reduces the effects of different sun angles and atmospheric effects (Coppin et al. 2004). The value of image regression as a change detection technique lies not only in its ability to identify changes in land-cover, but also in its analytic capability. Regression model can also be modified to specify relationships of many different forms.

Despite the preceding advantages, image regression techniques are slightly better than simple image subtraction (Ridd and Liu 1998). Since it is based on linearity assumption, this technique is not acceptable if a large proportion of the study area has changed between the two image dates. A key statistical assumption of regression is that relationships are stationary. Many years of geographic research have demonstrated that this is often not the case with spatial data (Cliff and Ord 1975; Haggett et al. 1977). Geographic data are typically non-stationary, and the spatial structure of data will affect the estimation of regression model parameters and, hence, residuals from the regression. Hanham and Spiker (2007) demonstrated a geographically weighted regression technique that accounts for non-stationary relationship in the estimation process; it also provides the researchers with an analytical tool to explore changes in the relationship between variables over space. This research has effectively used multi-temporal satellite imagery to detect and analyse the urban sprawl.

5.7 Principal Components Analysis Transformation

Principal components analysis (PCA) (Gonzalez and Wintz 1977; Faust 1989) is a multivariate statistical method for data summarisation and reduction. The fundamental assumption of PCA is the existence of some underlying structure between a set of physical variables that can be described by a smaller set of synthetic variables (Taylor 1977). PCA is used to reduce image dimensionality by defining new, uncorrelated bands composed of the principal components (PCs) of the input bands. PCs are computed by examining the correlation between input image bands, grouping highly correlated bands, and then calculating new bands that summarise the information contained in the original band set. Duda et al. (2001) may be referred for computational details.

The creation of new PCs can be considered as the translation and rotation of the original image axes (*eigenvectors*) in feature space to match patterns in the input variables. The first principal component (PC1) (having widest *eigenvalue*) accommodates the maximum possible variations of the input image. Each subsequent component is calculated such that it accounts for the maximum possible variance remaining, and is orthogonal to the axis of previous component. In the process of PCA, new PC bands are created by projecting the original data values in terms of the new axes (Jensen 2005; Liu and Mason 2009).

Urban change detection, using PCA, is based on the assumption that values of pixels of same location are constant through time if there is no change in land-cover. According to this condition, a high correlation among the pixels of same location will be found between different image dates. The most common method of utilizing PCA in change detection is to combine two images of different dates with n bands each into a single image with $2n$ bands; this image is then used for PCA and to compute PC images (Hardin et al. 2007). The transformation will produce $2n$ number of PC images. If the change area is limited or small, unchanged portion of the landscape will be represented by PC1, while successive components will represent areas of change (Fig. 5.4). If a change area is large or huge in amount then PC1 will highlight the changed areas. Different types of change will be represented by different PCs. (Byrne et al. 1980; Fung and LeDrew 1987). This technique is, however, sensitive to noise and temporal (seasonal/diurnal) illumination differences. It has been clearly demonstrated by Singh and Harrison (1985) that the use of

5.8 Change Vector Analysis

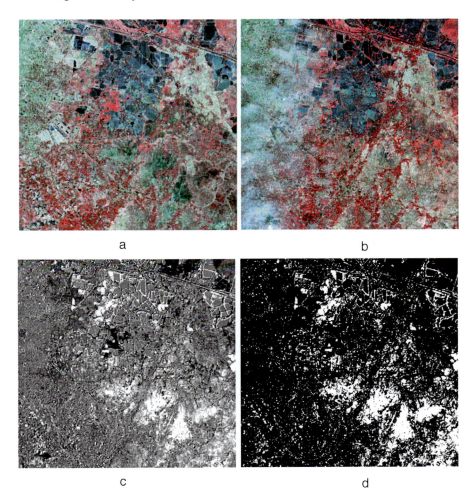

Fig. 5.4 (**a**) Infrared colour composite image of t_1; (**b**) Infrared colour composite image of t_2; (**c**) Second PC image; (**d**) Second PC image has been enhanced by changing the *brightness/contrast* to highlight the change

standardised variables (correlation matrix) in the analysis yields significantly different results.

5.8 Change Vector Analysis

Change vector analysis (CVA) is another method of mapping the changes in landcover (Chen et al. 2003a). Spectral information for a given pixel is represented in its *measurement vector*. The measurement vector of a pixel is the set of data file values for one pixel in all k bands. For example, for a given sensor having four spectral bands, the vector can be expressed as:

$$R = \begin{bmatrix} r_1 \\ r_2 \\ r_3 \\ r_4 \end{bmatrix} \quad (5.6)$$

Where, r_1 is the reflectance for the pixel in band 1, r_2 the reflectance in band 2, and so forth.

Given two dates of imagery, measurement vectors for the pixel of same location on two dates can be expressed as R_1 and R_2. Then the change vector can be defined as:

$$\Delta R = R_1 - R_2 = \begin{bmatrix} \Delta r_1 \\ \Delta r_2 \\ \Delta r_3 \\ \Delta r_4 \end{bmatrix} \quad (5.7)$$

Magnitude and *direction* (positive or negative) of change in vectors can be calculated by simple vector formulae (Malila 1980; Johnson and Kasischke 1998). This will result in a two band difference image where one band represents the magnitude of the change vectors and the other shows the direction of change. A simple formula to calculate the magnitude is (Chen et al. 2003a):

$$\| \Delta R \| = \sqrt{(\Delta r_1)^2 + (\Delta r_2)^2 + (\Delta r_3)^2 + (\Delta r_4)^2} \quad (5.8)$$

It represents the total grey-level difference between two dates. The greater the $\| \Delta R \|$ is, the higher is the possibility of change. A decision on change is made based on whether the change magnitude exceeds a specific threshold. Once a pixel is identified as change, the direction of ΔR can be examined further to determine the type of change.

However, CVA is also sensitive to noise and temporal (seasonal/diurnal) illumination differences, and challenged by selecting the threshold value that defines relevant change in magnitude and direction (similar to the image subtraction). Traditionally, the determination of threshold is empirical. Chen et al. (2003a) have shown how to determine this threshold in an effective way.

5.9 Artificial Neural Network

An *artificial neural network* (ANN), also called a *simulated neural network* (SNN) or commonly just *neural network* (NN), is an interconnected group of artificial *neurons* that uses a mathematical or computational model for information processing based on a connectionist approach to computation. In most cases, an ANN is an adaptive system that changes its structure based on external or internal information that flows through the network. In more practical terms ANNs are nonlinear statistical data modelling tools such as *regression* and *discriminant* analysis because they estimate functions adaptively rather than through some constrained mathematical algorithm like ordinary least squares. They can be used to model complex relationships between inputs and outputs or to find patterns in data.

Although the brain's computational method is very different from a computer (Haykin 1994), ANN tends to mimic the biologic brain's fault tolerance and learning capacity. According to Jensen and Hardin (2005), ANNs have been used in remote sensing applications to classify images (Bischof et al. 1992; Hardin 2000), and incorporate multisource data (Benediktsson et al. 1990). ANN classifiers have been successfully used with remote sensing data because they take advantage of the ability to incorporate non-normally distributed numerical and categorical GIS data and image spatial information (Jensen et al. 2000).

In supervised classification, *feed-forward back-propagation* ANNs can replace the typical maximum-likelihood classifier used to assign image pixels to a training class. Other types of ANN can serve as unsupervised classifiers or hybrid combinations of both strategies. For example, the *Learning Vector Quantizer* is a self-organising neural network map that can be adapted to either supervised or unsupervised learning strategies (Kohonen 1995).

According to Liu and Lathrop (2002), ANNs have the capability of detecting changes between temporal images, but the high cost of network training limits their widespread adoption. Because of this problem, the authors developed a method for training such ANNs that is quicker than traditional approaches while also facilitating efficient feature extraction. Two such examples of feed-forward back-propagation ANNs may be given from Liu and Lathrop (2002). The first network was trained with a stacked 12-band image created from the optical bands of the two temporal Landsat-TM images. The second network was trained with a six band PC set from the aforementioned 12-band image. When compared to the accuracy of the mapping produced by typical post-classification overlay, the results of the ANN were found superior. Further, the network trained with the PCA vectors produced better results than the stacked 12-band image.

Bruzzone et al. (1999) may be referred for a more advanced approach to change detection using feed-forward networks.

5.10 Decision Tree

Decision tree is another innovative method for urban change mapping (Im and Jensen 2005; Chan et al. 2001). Whereas ANNs were spawned by research into artificial intelligence, decision trees are used to represent rules in expert systems (systems nominally designed to model human expert knowledge and decision making strategies). Similar to ANNs, decision trees are also used for image classification. Decision trees have *nodes* that represent distinguishing attributes; and these nodes are decision points. 'Nodes in different levels of the tree are connected by *arcs*. Arcs exiting nodes represent different decisions made by examining the nodes' attribute state. To classify a pattern, a path is followed through the tree from root to leaf examining nodal attributes and making decisions along the way. A decision at a given node selects an arc and moves the decision to a leaf (representing a final class label) or to another node (requiring another conditional examination)' (Hardin et al. 2007).

Chan et al. (2001) compared relative effectiveness of decision trees with three other machine learning algorithms (i.e., multilayer ANN, maximum likelihood, learning vector quantizer) to map urban change on Tsing-yi Island, Hong Kong. They found that, for the binary change mapping, accuracy was 88, 77, 83, and 91% for the decision tree, maximum likelihood, ANN, and learning vector approaches respectively. However, Weismiller et al. (1977) state that a major problem with the decision tree approach is the complexity and large amount of computer core required to implement the algorithms. The method also requires apriori knowledge of the logical inter-relationship of the classes.

5.11 Intensity-Hue-Saturation Transformation

The *intensity-hue-saturation* (IHS) *transform* is a popular method for image fusion,[4] and yet represents another innovative method of detecting urban change between two satellite images. As described by Chen et al. (2003b), IHS change detection is based on an observation from image fusion i.e., when a forward and inverse IHS transform are performed for fusion purposes, a noticeable colour distortion will result if one of the images exhibits any spectral or textural differences related to temporal change.

This methods requires two images of different dates—one multispectral and other panchromatic. After initial rectification, as demonstrated by Chen et al. (2003b), contrast enhancement is to be performed on the panchromatic image to enhance urban-vegetation differences and high-pass filter to be applied for edge enhancement. Also they have enhanced the multispectral image using histogram equalisation. Three successive bands of multispectral image are to be transformed into IHS components. Then the intensity component should be replaced by the panchromatic image and retransformed to RGB space. This RGB image is then to be classified for the identification of change classes and no-change. The choice of original RGB bands to be transformed into IHS is a comparative trivial and chosen empirically.

---

[4] For more information on the IHS transform, see Harris et al. (1990) and Liu and Mason (2009). For an introduction to image fusion in remote sensing, see Pohl and Van Genderen (1998) and Liu and Mason (2009).

However, this method restricts the utilization of full spectral information contained within the multispectral/hyperspectral imageries that have more than three bands.

5.12 Econometric Panel

Kaufmann and Seto (2001) demonstrated how to detect urban change between multi-temporal dates, and to estimate the date of changes using *econometric panel* techniques. Their objective was to match yearly socioeconomic data with its concurrent annual land-cover change. For this approach, it was obvious to acquire satellite imageries of yearly interval and be compared; however, the researchers imagined that repetitive comparison of image pairs may result in unacceptable errors. Thus, to prevent the compounding of errors associated with post-classification comparison (discussed later), they have avoided repetitive comparison of image pairs.

In their study, Kaufmann and Seto (2001) have used nine images between 1988 and 1996. Images were classified into seven land-cover classes. In order to assign change dates to image pixels displaying change between 1988 and 1996 (and to avoid the pair-wise comparisons), they have adopted econometric panel techniques.

The process involves three steps: In the first step, estimation of regression equations for each of the bands for each of the land-cover classes. In the second step, the estimated regression equations for each class are used to calculate pixel values for change-classes and for each of the possible dates of change. In the third step, the date of change is identified by comparing the values of a pixel against the all possible dates of change using tests for predictive accuracy. The results of prediction accuracy were mixed and not always acceptable.

Important to mention that CVA, ANN, decision tree, IHS transform and econometric panel are innovative approaches to urban change detection, which have been introduced in recent years. All are more complicated than popular traditional methods. One of the research challenges faced by urban change detection practitioners is making these innovative approaches available to the broader user community; only then can the domain of their effectiveness be delimited (Hardin et al. 2007).

5.13 Image Classification and Post-classification Comparison

Image classification is the process of sorting pixels into finite number of individual classes or categories of data, based on their pixel values. Classification of remotely sensed data is used to assign corresponding levels with respect to groups with homogeneous characteristics, with the aim of discriminating multiple objects from each other within the image. The level is called class.

A human analyst attempting to classify features in an image uses the elements of visual interpretation to identify homogeneous groups of pixels, which represent various features or land-cover classes of interest. Digital image classification uses the spectral information represented by the digital numbers in one or more spectral bands, and attempts to classify each individual pixel based on this spectral information. This type of classification is termed 'spectral pattern recognition'. In either case, the objective is to assign all pixels in the image to particular classes or themes (e.g., water, coniferous forest, deciduous forest, corn, wheat, etc.). The resulting classified image is comprised of an array of pixels, each of which belongs to a particular theme, and is essentially a thematic 'map' of the original image scene.

There are many decision rules available for the image classification; for example, minimum distance to mean, maximum likelihood, linear discriminant, parallelepiped, feature space, one-pass clustering, sequential clustering, statistical clustering, K-means clustering, iterative self-organising clustering, RGB clustering, and so on. It is beyond the scope of this literature to discuss on image classification in detail (one may refer Lu and Weng 2007; Jensen 2005; ERDAS 2008; Bhatta 2008; Liu and Mason 2009).

In remote sensing it is often not easy to delineate the boundary between two different classes. For instance, there is transitive vegetation or mixed vegetation between forest and grassland. In such cases, as unclear defined class boundaries, fuzzy set theory can be usefully applied, in a qualitative sense (Wang et al. 2007). Several other advanced and innovative classification approaches are: neural network or fuzzy classification (Paola and Schowengerdt 1995; Gamba and Houshmand 2001; Zhang and Foody 1998; Gamba and Dell'Acqua 2003), spectral mixture analysis or subpixel classification (Alberti et al. 2004; Lu and Weng 2004; Yue et al. 2006; Xian and Crane 2005; Brown et al. 2000; Phinn et al. 2002), knowledge-based classification (ERDAS 2008), object-oriented or object-based classification (Doxani et al. 2008; Jacquin et al. 2008), etc. However, these are more complex, and often require additional data and exclusive understanding of image processing, and are computationally expensive. These approaches will have to overcome several hurdles before coming in force in practical applications. Most of the current applications rely on simple per-pixel classification of the imagery.

Post-classification comparison is currently the most popular method of urban change detection (Jensen et al. 1993) (Table 5.1). In post-classification comparison, each date of rectified imagery is independently classified to fit a common land-type schema (equal number and type of land-cover classes). The resulting land-cover maps are then overlaid and compared on a pixel-by-pixel basis. The result is a map of land-cover change (Fig. 5.5). This per-pixel comparison can also be summarised in a *'from-to' change matrix* (also called *transition matrix*) (Jensen 2005). The 'from-to' change matrix (Table 5.2) shows every possible land-cover change under the original classification schema and shows the areas of each change class (Howarth and Wickware 1981).

Various uses of change map and change matrix have made the post-classification comparison very popular. Its conceptual simplicity is another reason of its wide adaptation. Furthermore, in this approach, since each image is independently classified, atmospheric corrections are not necessary (Kawata et al. 1990; Song et al. 2001; Alphan 2003; Coppin et al. 2004). This method is also appropriate when

5.13 Image Classification and Post-classification Comparison

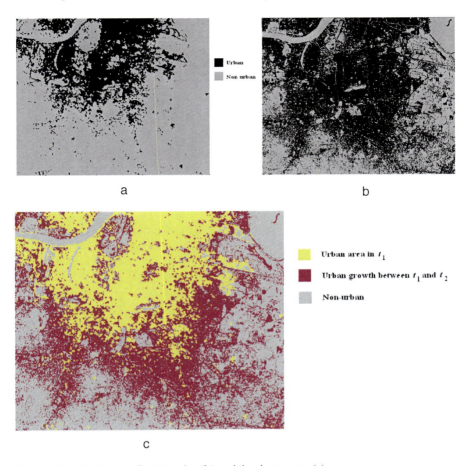

Fig. 5.5 Classified maps of t_1 (**a**) and t_2 (**b**), and the change map (**c**)

Table 5.2 One example of transition matrix between dates t_1 and t_2 (area in km^2). Unchanged areas occupy the diagonal of the matrix while changed areas are represented in the off-diagonal elements of the matrix

		Land-cover at time t_2						t_1 Total
		Class-A	Class-B	Class-C	Class-D	Class-E	Class-F	
Land-cover at time t_1	Class-A	**1,240**	3	1	3	7	0	1,254
	Class-B	8	**20,642**	2,058	297	2,383	4	25,392
	Class-C	12	592	**18,493**	226	1,641	15	20,979
	Class-D	0	3	52	**2,741**	18	0	2,814
	Class-E	13	86	959	44	**8,049**	2	9,153
	Class-F	1	1	0	0	3	**18,962**	18,967
	t_2 Total	1,274	21,327	21,563	3,311	12,101	18,983	

imageries of different dates have substantially different sun or look angles, or come from different sensors (Yang and Lo 2002).

Despite its wide adaptation, post-classification comparison is error prone (Hardin et al. 2007). The accuracy of the change map is highly dependent upon the accuracy of the independently classified maps (Aspinall and Hill 1997; Serra et al. 2003; Yuan et al. 2005). All of the errors within the independently classified maps will be transferred in the change map. For example, if two land-cover maps each have a producer's accuracy of 90%, the accuracy of the post-classification change map accuracy will be about 81% (0.9 × 0.9 × 100) (Stow et al. 1980).

5.14 Challenges and Constraints

It is important to mention that urban change detection is challenged by several factors. Changes in apparent reflectance can be caused by many factors other than the land-cover conversion. These factors include (Riordan 1980; Ingram et al. 1981; Du et al. 2002; Prenzel 2004; Paolini et al. 2006; Hardin et al. 2007):

- Atmospheric attenuation—different atmospheric condition and thereby different illumination in different image dates
- Misregistration—registration error between multiple image dates
- Topographic attenuation—different shadow effects on the images of different dates
- Different phenological[5] stage and/or seasonal variability—change in appearance of same object (e.g., vegetation) between image dates
- Sensor spatial, spectral and radiometric resolution differences—images of different sensors in use
- Changes in sensor response for same sensor due to drift or age
- Changes in viewing and/or sun azimuth—difference in shadow effects
- Changes in sun elevation—different illumination and shadow effects.

Of these aforementioned error sources, atmospheric differences and misregistration are the most accountable in change detection (Coppin and Bauer 1994). Apparent differences due to changes in sun azimuth and phenology can often be eliminated or reduced to ignorable levels by using imagery collected by the same sensor on anniversary dates (Jensen 2005). Unfortunately, collection of cloud free imagery on anniversary dates is not assured.

Prenzel (2004) recommended the following ideal conditions for remote sensing data that are preferred for systematic mapping of land-cover changes:

---

[5] Phenology is the study of periodic plant (and animal) life cycle events and how these are influenced by seasonal and inter-annual variations in climate.

5.14 Challenges and Constraints

- Data collected by the same sensor will have similar characteristics (e.g. spatial, spectral and radiometric resolution; geometry, radiometric response).
- Data acquired under clear atmospheric conditions (i.e. free of cloud and haze) will assist in the identification of real change on the ground surface.
- Data collected on anniversary dates will have similar surface conditions for the area under study (i.e. consistent plant phenology) as well as consistent sun-terrain-sensor geometry.

Chapter 6
Measurement and Analysis of Urban Growth

6.1 Introduction

The process of mapping of urban growth results in the creation of abstracted and highly-simplified change maps of the study area (as shown in Fig. 5.5). Examining these thematic change maps, even cursorily, one may see that expansion of built-up has different signatures: some areas are very compact while in others more open space between built-up areas. In some of the areas the boundary between the built-up and non-built-up is rather sharp, while in others these classes dissolve into each other. One can also see the infill of the open spaces between already built-up areas that results in their consolidation; or, one can understand whether the city is becoming more monocentric or polycentric over time. Surely, one can grasp these patterns intuitively, but they fall short of providing solid evidences for debating and deciding upon the future. To describe these different patterns intelligently, to understand how they change over time, to compare one subpart with others, or to explain the variations among these patterns statistically, we need to select quantitative measures that summarise one or another of their properties. Recently, urban change detection focus has been shifted from detection to quantification of change, measurement of pattern, and analysis of pattern and process of urban growth and sprawl.

This chapter attempts to shed some light on how the urban growth can be measured and analysed. The entire discussion is divided into three main sections—*transition matrices*, *spatial metrics*, and *spatial statistics*. However, important to realise, these three are rarely independent, rather often overlapping in many urban growth analysis applications.

> It is worth mentioning that measurement and analysis of urban growth and the process of sprawl require at least two temporal dataset; however, sprawl can be measured or analysed as a pattern using the data from a single date. This single data can be used to study and analyse the urban form that can effectively be used to characterise the urban sprawl.

6.2 Transition Matrices

Transition matrix is a table (as shown in Table 5.2) that allows the user to measure changes among different land-cover/land-use classes over a time period (Jensen 2005). It is focused on observing changes in landscape and the mechanisms behind that change. Remotely sensed imageries can be classified into different land-cover types.[1] By using the classified data from two different dates, transition matrix can be constructed to determine the area in each land-cover class which has changed to each other land-cover class.

Geographic information system (GIS) tools allow these changes to be quantified from classified remote sensing data in space and time to show the spatial pattern and composition of land-cover in a dynamic representation. These changes can be integrated with social and biophysical data to determine the factors behind the land-cover change process or the consequences of such transition. Often these linear or nonlinear relationships can be modelled mathematically and statistically. Transition matrices are also useful for validating the preceding models.

Transition matrix is perhaps the most simple and widely used technique to describe and understand the changes in landscape and to model them with the causes and consequences of such changes. The main advantage of using transition matrix is the sensor independence. Since it does not require any mathematical or statistical operations among the images, one may consider multi-temporal images of different sensors. Resampling of spatial resolution to make the pixels of same size from different sensors is not required because it aggregates or summarises the data for a geographic region or sub-region. However, it is important to realise that aggregation of geographic data looses its spatial detail; and instead of each and every geographic location the data represents a geographic region or sub-region.

6.2.1 Transition Matrix in Urban Growth Analysis

Many of the researchers prefer to present the urban change data in a transition matrix for general understanding. Although other methods are more preferred for the analysis of urban growth and sprawl, transition matrix is most preferred to measure the urban growth or to explain the transition among land-cover classes over a geographic area. However, owing to limitations of spatial analysis, focus on transition matrices is generally overlooked. Despite their limitations often they became the basis of rigorous analysis using descriptive statistics and metrics (e.g., Angel et al. 2005; Almeida et al. 2005; Batisani and Yarnal 2009; Bhatta 2009a; Bhatta et al. 2010a; Dewan and Yamaguchi 2009; Aguayo et al. 2007).

---

[1] Land-use details are not directly observable by a remote sensor; inferences about land-use can only be maid by the experts. Therefore, analysis of urban growth, from remote sensing data, is mainly focused on land-cover, rather than land-use.

6.3 Spatial Metrics

Recently, innovative approaches to urban land-use planning and management such as sustainable development and smart growth have been proposed and widely being discussed. However, most of the implementations rely strongly upon available information and knowledge about the causes, chronology, and effects of urban change processes (Herold et al. 2005a). Despite the recent proliferation of new sources of remote sensing data and tools for data processing and analysis, these have not directly led to an improved understanding of urban phenomena.

This section explores how remote sensing data and tools for processing and analysis in combination with *spatial metrics* can improve the understanding of urban spatial structure and change processes, and can support the modelling of these processes. Spatial metrics are numeric measurements that quantify spatial patterning of land-cover patches,[2] land-cover classes, or entire landscape mosaics of a geographic area (McGarigal and Marks 1995). These metrics have long been used in *landscape ecology*[3] (where they are known as *landscape metrics* (Gustafson 1998; Turner et al. 2001)) to describe the ecologically important relationships such as connectivity and adjacency of habitat reservoirs. Applied to fields of research outside landscape ecology and across different kinds of environments (in particular, urban areas), the approaches and assumptions of landscape metrics may be more generally referred to as *spatial metrics* (Herold et al. 2005a). Spatial or landscape metrics, in general, can be defined as quantitative indices to describe structures and patterns of a landscape (O'Neill et al. 1988). Herold et al. (2005a) defined it as 'measurements derived from the digital analysis of thematic-categorical maps exhibiting spatial heterogeneity at a specific scale and resolution'.

The analysis of spatial structures and patterns are central to many geographic researches. Spatial primitives such as location, distance, direction, orientation, linkage, and pattern have been discussed as general spatial concepts in geography (Golledge 1995) and they have been implemented in a variety of different ways.

Under the name of landscape metrics, spatial metrics are already widely used to understand and quantify the shape and pattern of vegetation in natural landscapes (O'Neill et al. 1988; Gustafson 1998; Hargis et al. 1998; McGarigal et al.

---

[2] 'Patch' is a term fundamental to landscape ecology; it is a relatively homogeneous area that differs from its surroundings. Patches are the basic unit of the landscape that change and fluctuate; this process is called *patch dynamics*. The concept of a patch, in general, is intuitive; we all seem to understand what constitutes a patch. However, a clear definition of patch can be given as 'a nonlinear surface area differing in appearance from its surroundings' (Forman and Godron 1986). Converting this definition into computer algorithm to identify patches on a remote sensing image 'a contiguous group of pixels of the same land-cover category'. What does it mean by 'contiguous' is well explained by Turner et al. (2001).

[3] Landscape ecology is the science of studying and improving the relationship between spatial pattern and ecological processes on a multitude of landscape scales and organisational levels.

2002). Landscape metrics are used to numerically explain spatial structure of landscape or *landscape structure*.[4] These metrics are useful to understand how structures affect system interactions in a heterogeneous landscape, numerical comparison of landscapes, and the recognition and monitoring of landscape change (Turner 1989; O'Neill et al. 1999; Turner et al. 2001). Furthermore, quantification of landscape structure enables the scientific transition from an inductive to deductive logic model wherein hypotheses can be formed and tested (Curran 1987; Dietzel et al. 2005). Shift in environmental ecology from a qualitative to quantitative basis was mainly fueled by metrics developed to quantify natural landscape structure and pattern (Hobbs 1999).

Metrics and methods used in landscape ecology are often influenced by others such as *computational complexity theory*,[5] *fractal geometry*,[6] and *spatial statistics*. Commonly used metrics can be subdivided into two broad categories (Hardin et al. 2007):

- Measurement of individual patch characteristics (e.g., size, shape, perimeter, perimeter-area ratio, fractal dimension).
- Measurement of whole landscape characteristics (e.g., richness, evenness, dispersion, contagion). Metrics of landscape characteristics are typically more computationally and analytically complex than individual patch metrics (Farina 1998).

Landscape metrics have found important applications in quantifying urban growth, sprawl, and fragmentation (Hardin et al. 2007). Herold et al. (2002) showed an early landmark in this shift by establishing that low-density residential, high density residential, and commercial zones can be discriminated by spatial metrics such as fractal dimension, landscape percentage, patch density, edge density, patch size standard deviation, contagion index, and area weighted average patch fractal dimension. These metrics were also capable of quantifying the land conversion. When applied to multi-scale or multi-temporal datasets, spatial metrics can be used to analyse and describe change in the degree of spatial heterogeneity (Dunn et al. 1991; Wu et al. 2000).

6.3.1 Remote Sensing, Spatial Metrics, and Urban Modelling

Herold et al. (2005a) illustrated a simple conceptual framework (Fig. 6.1), consisting of three main components: remote sensing, spatial metrics and urban modelling,

---

[4] Spatial structure of landscape (or landscape structure) means the size, shape, area, composition, number, and position of ecosystems that make up a landscape.

[5] Complexity theory focuses on classifying problems according to their inherent difficulty (Allen 1997; Pooyandeh et al. 2007; Batty 2009).

[6] Fractal geometry is a rough or fragmented geometric shape that can be split into parts, each of which is (at least approximately) a reduced-size copy of the whole (Mandelbrot 1982; Lam and Lee 1993).

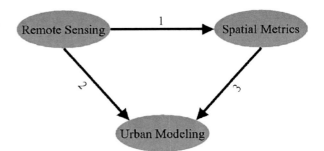

Fig. 6.1 General framework for analysis and modelling of spatial urban dynamics (Herold et al. 2005a)

and their interrelations. While the potential direct contribution of remote sensing to urban modelling is fairly well understood (relationship 1 in Fig. 6.1), but the combined use of remote sensing and spatial metrics will lead to new levels of understanding of how urban areas grow and change (relationships 2 and 3 in Fig. 6.1), namely the pattern and process of urban growth.

In recent years, the use of computer-based models for urban growth analysis has greatly increased, and they have shown the potentials to become important tools for urban planning and management. This development was possible by the increased data resources, improved usability of multiple spatial datasets and tools for their processing, as well as an increased acceptance of models (Klosterman 1999; Sui 1998; Wegener 1994). However, the application and performance of urban models are strongly depend on the quality and scope of the data available for parameterisation, calibration and validation, as well as the level of understanding built into the representation of the processes being modelled (Batty and Howes 2001; Longley and Mesev 2000). Remote sensing data products have often been incorporated into urban modelling applications as additional sources of spatial data (relationship 1 in Fig. 6.1), primarily for historical land-use/land-cover information (Acevedo et al. 1996; Clarke et al. 2002; Meaille and Wald 1990). Relationship 3 (Fig. 6.1) corresponds to the use of spatial metrics in urban modelling. This relationship has been proposed in a few studies that use spatial metrics to refine and improve remote sensing data for urban models, for model calibration and validation, or in studies of urban landscape heterogeneity and dynamic change processes (Alberti and Waddell 2000; Herold et al. 2003a; Parker et al. 2001).

6.3.2 Spatial Metrics in Urban Growth Analysis

Based on the work of O'Neill et al. (1988), sets of different spatial metrics have been developed, modified and tested (Hargis et al. 1998; McGarigal et al. 2002; Riitters et al. 1995). Many of these quantitative measures are implemented in the public domain statistical package FRAGSTATS[7] (McGarigal et al. 2002). Spatial metrics can be grouped into three broad classes: *patch*, *class*, and *landscape* metrics.

---

[7] Available at http://www.umass.edu/landeco/research/fragstats/fragstats.html

Patch metrics are computed for every patch in the landscape, *class metrics* are computed for every class in the landscape, and *landscape metrics* are computed for entire patch mosaic. There are numerous types of spatial metrics that are found in the existing literature, for example: area/density/edge metrics (patch area, patch perimeter, class area, number of patches, patch density, total edge, edge density, landscape shape index, largest patch index, patch area distribution); shape metrics (perimeter-area ratio, shape index, fractal dimension index, linearity index, perimeter-area fractal dimension, core area metrics (core area, number of core areas, core area index, number of disjunct core areas, disjunct core area density, core area distribution); isolation/proximity metrics (proximity index, similarity index, proximity index distribution, similarity index distribution); contrast metrics (edge contrast index, contrast-weighted edge density, total edge contrast index, edge contrast index distribution); contagion/interspersion metrics (percentage of like adjacencies, clumpiness index, aggregation index, interspersion & juxtaposition index, mass fractal dimension, landscape division index, splitting index, effective mesh size); connectivity metrics (patch cohesion index, connectance index, traversability index); and diversity metrics (patch richness, patch richness density, relative patch richness, Shannon's diversity index, Simpson's diversity index, Shannon's evenness index, Simpson's evenness index). One may refer the manual of FRAGSTATS for detailed discussion.

Recently there has been an increasing interest in applying spatial metrics techniques in urban environments because these help to bring out the spatial component in urban structure (both intra- and inter-city) and in the dynamics of change and growth processes (Alberti and Waddell 2000; Barnsley and Barr 1997; Herold et al. 2002). Herold et al. (2005a) argued that the combined application of remote sensing and spatial metrics can provide more spatially consistent and detailed information on urban structure and change than either of these approaches used independently.

Based on some studies, Parker et al. (2001) summarise the usefulness of spatial metrics with respect to a variety of urban models and argue for the contribution of spatial metrics in helping link economic processes and patterns of land-use. They have used an agent-based model of economic land-use for decision-making, which has resulted in specific theoretical land-use patterns. They conclude that urban landscape composition and pattern, as quantified with spatial metrics, are critical independent measures of the economic landscape function and can be used for an improved representation of spatial urban characteristics and for the interpretation and evaluation of modelling results. Alberti and Waddell (2000) substantiate the importance of spatial metrics in urban modelling.

Schneider et al. (2005) also discuss the recent shift in focus from urban change detection to change quantification. Their statistics highlighted the effect of economical, social, and government policy forcing mechanisms on urban structure. As illustrated by Schneider et al. (2005) and articulated by Seto and Fragkias (2005), the quantification of urban growth with landscape metrics represents a significant enhancement to the calculation of yearly land-type acreage changes. Although area figures enable change rate calculations, spatial metrics make possible the evaluation of changing urban spatial pattern—an important additional piece of information for planners seeking to control urban growth (Hardin et al. 2007).

Seto and Fragkias (2005) used spatial metrics to quantify change in four cities over an 11 year period (1988–1999). Using satellite imagery, maps of change for several years were constructed. Urban growth rates were then annualised, and using the annualised change images, six spatial landscape metrics were calculated for three concentric buffer zones centred on each of the four cities. The metrics selected were intended to describe urban form complexity and size, and included total urban area, edge density, urban patch count, mean patch fractal dimension, average patch size, and patch size coefficient of variation. Calculation of urban change rates during the ten year period was also done. As detailed by Seto and Fragkias, key aspects of urban development in the two cities were illuminated by the metrics. Envelopment and multiple nuclei growth were revealed as the primary urban expansion processes. Changing administrative practices to control (or not) land-use development were likewise reflected in the metrics. These results illustrate that investigating temporal urban change via landscape metrics is a valuable procedure both to quantify change and link its spatial pattern to cultural practices and processes. Yu and Ng (2006) have also demonstrated a similar study.

Yu and Ng (2007) have employed landscape metrics in addition to gradient analysis on remote sensing data to analyse and compare both the spatial and temporal dynamics of urban sprawl in Guangzhou, China. They have considered four temporal images, and to detect the dynamics of landscape pattern two transects were selected that cut across the entire Guangzhou city. Landscape metrics were then calculated for each block at the class and landscape levels using the FRAGSTATS software. Eight metrics were selected in this study, which can fully reflect their conceptual basis and reduce correlation and redundancy. The results show that the combination of gradient analysis and landscape metrics can characterise the complex spatial pattern of urban growth.

Herold et al. (2005a) coined the term 'spatial metric growth signatures' to describe this use of landscape metrics. Metrics were not only used to describe historical structure and predicted future urban form, but were also competent to measure goodness of fit between an urban growth model and historical data. The authors also illustrated the use of spatial metrics as a means of visualising urban modelling results.

Huang et al. (2007) have considered seven spatial metrics (compactness, centrality, complexity, porosity, and density) to analyse the urban form of 77 metropolitan areas in Asia, US, Europe, Latin America and Australia. Comparison of the spatial metrics was made between developed and developing countries, and then among world regions. A cluster analysis was used to classify the cities into groups in terms of these spatial metrics. They also explored the origins of differences in urban form through comparison with socioeconomic developmental indicators and historical trajectories in urban development. The result clearly demonstrates that urban agglomerations of developing world are less sprawling and dense than their counterparts in either Europe or North America. However, they have not considered temporal images, and therefore it lacks the insight of urban growth or sprawl process.

Martinuzzi et al. (2007) also considered a single temporal image to identify the low-density and high-density urban development in addition to rural development by applying a moving window in conjunction with census data. They have used textural index for the said purpose. They have also developed a classification schema to categorise the relative tendency to sprawl of urban developments.

Other studies include urban gradient analysis (Luck and Wu 2002) and the assessment of forest fragmentation (Civco et al. 2002).

6.4 Spatial Statistics

Spatial statistics is concerned with the spatially referenced data and associated statistical models and processes (NRC 1991). In statistics, spatial statistics includes any form of formal techniques which study entities using their topological, geometric, or geographic properties. The phrase actually refers to a variety of techniques; many of which still in their early development. The phrase is often used in a more restricted sense to describe techniques applied to structures at the human scale, most notably in the analysis of geographic data. The phrase is even sometimes used to refer to a specific technique in a single area of research, for example, to describe geostatistics. NRC (1991) illustrates a number of spatial statistics in several application areas.

Turner et al. (2001) have identified two main applications of spatial statistics: (1) identifying the spatial scales over which patterns (or processes) remain constant (or, alternatively, the scales at which significant changes in pattern and process can be detected) and (2) interpolating or extrapolating point data to infer the spatial distributions of variables of interest. Páez and Scott (2004) have reviewed several spatial statistics for urban analysis. They have addressed several major analytical issues, along with techniques to deal with them.

6.4.1 Types of Spatial Statistics

There are a number of spatial statistics in the existing literature which are often difficult to categorise. Spatial data comes in many varieties and it is not easy to arrive at a system of classification that is simultaneously exclusive, exhaustive, imaginative, and satisfying (Graham and Fingelton 1985). However, this section makes an attempt to categorise different types of spatial statistics.

6.4.1.1 Spatial Autocorrelation

Spatial autocorrelation measures and analyses the degree of dependency among observations in a geographical space (Fig. 6.2). Classic spatial autocorrelation statistics (such as *Moran's I* and *Geary's C*) require measuring a spatial weights matrix

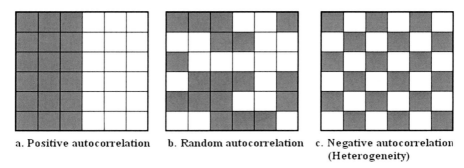

a. Positive autocorrelation   b. Random autocorrelation   c. Negative autocorrelation
                                                              (Heterogeneity)

Fig. 6.2 Spatial autocorrelation

that reflects the intensity of the geospatial relationships between observations in a neighbourhood, e.g., the distances between neighbours, the lengths of shared border, or whether they fall into a specified directional class, etc. Classic spatial autocorrelation statistics compare the spatial weights to the covariance relationship at pairs of locations. Spatial autocorrelation that is more positive than expected from random indicate the clustering of similar values across geographical space, while significant negative spatial autocorrelation indicates that neighbouring values are more dissimilar than expected by chance, suggesting a spatial pattern similar to a chess board.

Spatial association is one of the major concerns in the analysis of spatial autocorrelation for spatial data, is the tendency of variables to display some degree of systematic spatial variation. In urban studies, this often means that high values are found near other high values and low values appear in geographical proximity (Páez and Scott 2004). Sometimes, however, the effect may have a negative quality when the ordering reflects systematic dissimilarity among neighbouring observations. Spatial association may be caused by a variety of spatial processes, including, among others, interaction, exchange and transfer, and diffusion and dispersion. It can also result from missing variables and unobservable measurement errors in multivariate analysis. Two exploratory techniques of spatial association are *local analysis* of spatial association (distance-based statistic; Getis and Ord 1993; Ord and Getis 1995) and *global statistic* of spatial association (Anselin 1995).

Classic spatial autocorrelation statistics (such as *Moran's I* and *Geary's C*) are global in the sense that they estimate the overall degree of spatial autocorrelation for a dataset. Local spatial autocorrelation statistics provide estimates disaggregated to the level of the spatial analysis units, allowing assessment of the dependency relationships across space.

6.4.1.2 Spatial Heterogeneity

The possibility of *spatial heterogeneity* suggests that the estimated degree of autocorrelation may vary significantly across geographical space. Urban processes often exhibit patterns of spatial heterogeneity—that is, they do not always operate in exactly the same way over space. Spatial heterogeneity is frequently thought to result from large-scale regional effects or administrative subdivisions that delimitate the reach of some processes (e.g., zoning). In statistical terms, spatial heterogeneity can be represented as structural variation in the definition of the variance or as systematic variation in the mean of the process. Three popular models that deal with spatial heterogeneity are *switching regressions* (Quandt 1958), *multilevel models* (Jones 1991; Duncan and Jones 2000) and *geographically weighted regression* (GWR) (Brunsdon et al. 1996, 1999). Of these methods, the first two constitute a compromise between global-local modelling. GWR on the other hand is a local form of spatial statistics.

Spatial regression methods capture spatial dependency in regression analysis, avoiding statistical problems such as unstable parameters and unreliable significance tests, as well as providing information on spatial relationships among the variables involved. Depending on the specific technique, spatial dependency can enter the regression model as relationships between the independent variables and the dependent, between the dependent variables and a spatial lag of itself, or in the error terms. GWR is a local version of spatial regression that generates parameters disaggregated by the spatial units of analysis. This allows assessment of the spatial heterogeneity in the estimated relationships between the independent and dependent variables.

6.4.1.3 Spatial Interpolation

Spatial interpolation methods estimate the variables at unobserved locations in geographical space based on the values at observed locations (Liu and Mason 2009). Spatial autocorrelation is assumed so that an unknown value can be estimated from the neighbourhood of values. The aim is to create a surface that models the sampled phenomenon so that the predicted values resemble the actual ones as closely as possible. Basic methods include *inverse distance weighting* (that attenuates the variable with decreasing proximity from the observed location) and *kriging* (that interpolates across space according to a spatial lag relationship that has both systematic and random components). Kriging is more sophisticated than inverse distance weighting and can accommodate a wide range of spatial relationships for the unknown values between observed locations. Kriging provides optimal estimates given the hypothesised lag relationship, and error estimates can be mapped to determine if spatial patterns exist.

6.4.1.4 Spatial Interaction

Spatial interaction estimates the flow of people, material or information between locations in geographical space. Factors can include origin propulsive variables such as the number of commuters in residential areas, destination attractiveness variables such as the amount of office space in employment areas, and proximity relationships between the locations measured in terms such as driving distance or travel time. In addition, the topological or connective relationships between areas must be identified, particularly considering the often conflicting relationships between distance and topology; for example, two spatially close neighbourhoods may not display any significant interaction if they are separated by a highway.

After specifying the functional forms of these relationships, the analyst can estimate model parameters using observed flow data and standard estimation techniques such as ordinary least squares or maximum likelihood. Competing destinations versions of spatial interaction models include the proximity among the destinations (or origins) in addition to the origin-destination proximity; this captures the effects of destination (origin) clustering on flows. Computational methods such as artificial neural networks can also estimate spatial interaction relationships among locations and can handle noisy and qualitative data. Fotheringham and O'Kelly (1989) discussed a number of spatial interaction models and their applications.

6.4.1.5 Modelling and Simulation

Spatial interaction models are *aggregate* and *top-down*: they specify an overall governing relationship for flow between locations. This characteristic is also shared by urban models such as those based on mathematical programming, flows among economic sectors, or bid-rent theory. An alternative modelling perspective is to represent the system at the highest possible level of desegregation and study the *bottom-up* emergence of complex patterns and relationships from behaviour and interactions at the individual level.

Complex adaptive systems theory as applied to spatial statistics suggests that simple interactions among proximal entities can lead to intricate, persistent and functional spatial entities at aggregate levels. Two fundamentally spatial simulation methods are *cellular automata* (Benenson and Torrens 2004; White and Engelen 1997) and *agent-based modelling* (Parker et al. 2003) (refer Chap. 7). Cellular automata modelling imposes a fixed spatial framework such as grid cells and specifies rules that dictate the state of a cell based on the states of its neighbouring cells. With the progress of time, spatial processes emerge as cell's change of state based on its neighbours; this alters the conditions for future time periods. For example, cells can represent locations in an urban area and their states can be different types of land-use. Patterns that can emerge from the simple interactions of local land-uses include office districts and urban sprawl. Agent-based modelling uses software entities (*agents*) that have purposeful behaviour (*goals*) and can react, interact and modify their environment while seeking their objectives. Unlike the cells in cellular automata, agents can be mobile with respect to space. For example, one could model traffic flow and dynamics using agents representing individual vehicles that try to minimize travel time between specified origins and destinations. While pursuing minimal travel times, the agents must avoid collisions with other vehicles also seeking to minimise their travel times. Cellular automata and agent-based modelling are divergent yet complementary modelling strategies. They can be integrated into a common geographic automata system where some agents are fixed while others are mobile (refer Chap. 7).

6.4.2 Spatial Statistics in Urban Growth Analysis

A number of references for the applications of spatial statistics in urban analysis or urban systems analysis can be given from the existing literature (e.g., Ortúzar and Willumsen 2001; Meyer and Miller 2001; Miller and Shaw 2001; Bolduc et al. 1995; Kwan 2000; Bhat and Zhao 2002; Bennion and O'Neill 1994; You et al. 1997a,b; Ding 1998; Horner and Murray 2002; Steenberghen et al. 2004).

Studies that relate the urban growth include, for example, analysis of relation of population density to distance from a central business district (CBD). According to the theory of monocentric city the population density should decrease with distance from the centre of the city, a function commonly used to describe this effect is the negative exponential, which is closely related to the relatively simple analytical framework required by economic theory. It has been noted, however, that the exponential function is not sophisticated enough to explain the spatial variability observed in many empirical situations (Páez and Scott 2004). A number of researchers have applied spatial statistics to address this problem. Bender and Hwang (1985), for example, developed a switching regression that takes into account variations with distance from the CBD. Alperovich and Deutsch (2002) proposed a more flexible switching regression specification that controls for directional variation in distance-decay. Switching regressions can identify heterogeneous regimes of population density or other attributes, such as land prices, in a city. Baumont et al. (2004) have used autoregressive models to study population and employment variations. Páez et al. (2001) used spatially switching-spatially autoregressive models to explore land price variation in a city. McMillen (2001) have shown a different approach to identify sub-centres of a polycentric city that uses GWR. Hanham et al. (2009) have used GWR to investigate spatially varying relationship between local land-use policies and urban growth. Their research examines the geography of urban growth and public policy using Landsat data. They focus on observed land-use change and the relationship between change and land-use policy.

Other than the aforementioned examples, autoregressive models have been used to analyse housing prices and neighbourhood effects (Tse 2002) and the effect of transportation infrastructure on housing prices (Haider and Miller 2000). Páez and Suzuki (2001) applied a *dynamic spatial logit model* to investigate the neighbourhood effects on land-use change. Their findings suggest that residential or commercial construction tends to encourage the same type of development in neighbouring zones. Another example of spatial statistics is study of Kanaroglou et al. (2002) who have demonstrated spatial interpolation techniques to derive urban pollution maps.

However, important to mention that many spatial statistical analyses are not explicitly dependent on remote sensing data. They require addition data or even they can run without the support of remote sensing.

> Spatial metrics vs. spatial statistics:
>
> Spatial statistics allow the quantification of spatial structure from sampled data; where as spatial metrics characterise the geometric and spatial properties of mapped data (e.g., mosaic of patches). Spatial statistics describe the degree of spatial autocorrelation, that is, the spatial dependency of the values of a variable (or self-correlation) that has been sampled at various geographical coordinates. Usually, such samples are gathered to better understand the spatial heterogeneity of data. Such quantitative knowledge about the spatial structure of the data can then be used to group samples into relatively spatially homogeneous clusters (i.e., patches) in the form of spatial metrics. Spatial statistics use point data for some property that is assumed to be spatially continuous across the landscape. Thus, they do not require categorisation of the landscape nor do they assume a patchy structure or the presence of boundaries (Turner et al. 2001). Gustafson (1998) notes that the categorical and point-data approaches to the description of spatial heterogeneity are seldom combined in most studies, leading to the appearance that they are not complementary. However, each approach offers advantages and disadvantages that depend both on the question being asked and the nature of the system being measured (Turner et al. 2001).

6.5 Quantification and Characterisation of Sprawl

Quantifying the urban growth from remote sensing data is not a difficult task; however, quantification of sprawl, as a pattern or process, is a real challenging issue. Wilson et al. (2003) said that without a universal definition, quantifying and modelling of urban sprawl is extremely difficult. Creating an urban growth model instead of an urban sprawl model allows us to quantify the amount of land that has changed to urban uses, and lets the user decide what he or she considers as urban sprawl. Angel et al. (2007) also support this concept. This statement, however, discourages the quantification efforts and makes the sprawl phenomenon more ambiguous. 'Although there have been many studies on the measurement of urban form they have limitations in capturing the characteristics of urban sprawl' (Yeh and Li 2001a).

There are many researchers who have shown interests in quantifying the urban sprawl. Worth mentioning that sprawl can be measured in relative and absolute scales. Absolute measurements are capable to create a black-and-white distinction between a sprawled city and a compact city. Relative measures, in contrast, quantify several attributes that can be compared between cities. In this case, whether

98   6 Measurement and Analysis of Urban Growth

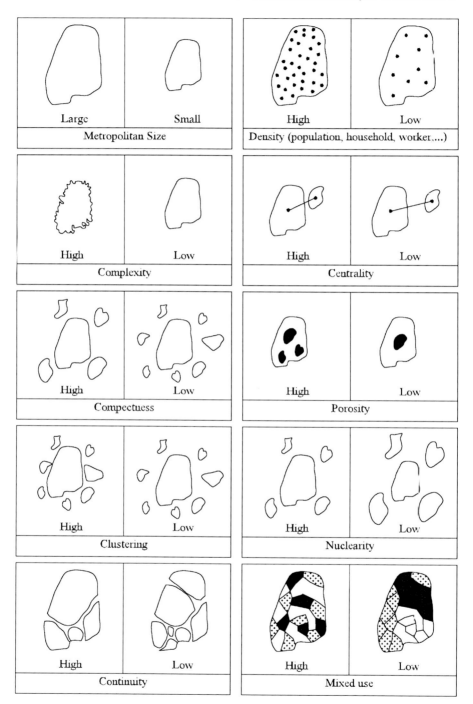

Fig. 6.3 Schematic diagrams of sprawl quantification and characterisation dimensions

6.5 Quantification and Characterisation of Sprawl

the city is sprawled or not is generally decided by the analyst, or even left without characterising the sprawl. It is important to mention that most of the sprawl measurement techniques, in general, relative measures; and can be used as indicators of sprawl. Absolute measurement of sprawl is never possible with these measures unless we define a threshold towards the black-and-white characterisation of sprawling and non-sprawling. Figure 6.3 illustrates some sprawl-characterisation dimensions schematically.

The review of literature has found that quantification of sprawl, either in relative or in absolute scale, has used all of three (transition matrix, spatial metrics, and spatial statistics) urban growth measurement and analysis approaches. In some of the instances, these approaches are overlapping and therefore difficult to discuss under a specific class. Moving away from tools to characteristics, Tsai (2005) has classified the urban sprawl studies into three classes—*density*, *diversity* and *spatial-structure pattern*. However, density and diversity may also refer to spatial structure, such as, built-up density or patch density, and land-cover diversity. Therefore, distinct classification may not be possible in general.

Galster et al. (2001) identified eight conceptual dimensions of land-use patterns as sprawl (Table 6.1). These dimensions are *density, continuity, concentration, clustering, centrality, nuclearity, mixed uses,* and *proximity*. Under the name of *sprawl metrics*, Angel et al. (2007) have demonstrated five metrics for measuring manifestations of sprawl (Table 6.2) and five attributes for characterising the sprawl (Table 6.3). Under each attribute they have used several metrics to measure the sprawl phenomenon. However, they have not mentioned any threshold that could be used for distinguishing a sprawling city from a non-sprawling city. Furthermore, interpretation of results from these metrics is also difficult and confusing since metrics are huge in number and one may contradict with other (Bhatta et al. 2010b).

Table 6.1 Dimensions of sprawl (Galster et al. 2001)

Dimension	Definition
Density	The average number of residential units per square mile of developable land[a]
Continuity	The degree to which developable land has been built upon at urban densities in an unbroken fashion
Concentration	The degree to which development is located disproportionately rather than spread evenly
Clustering	The degree to which development has been tightly bunched to minimize amount of land occupied by residential or non-residential uses
Centrality	The degree to which residential and/or non-residential development is located close to a central business district
Nuclearity	The extent to which an area is characterised by a mononuclear development pattern (as opposed to polynuclear)
Mixed uses	The degree to which multiple land-uses exist within the same small area and its prevalence throughout a region
Proximity	The degree to which different land-uses are close to each other

[a] Where new construction can be made.

Table 6.2 Metrics for measuring manifestations of sprawl (Angel et al. 2007)

Metric	Definition
Main urban core	The largest contiguous group of built-up pixels in which at least 50% of the surrounding neighbourhood is built-up
Secondary urban core	Built-up pixels not belonging to the main urban core that have neighbourhoods that are at least 50% built-up
Urban fringe	Built-up pixels that have neighbourhoods that are 30–50% built-up
Ribbon development	Semi-contiguous strands of built-up pixels that are less than 100 m wide and have neighbourhoods that are less than 30% built-up
Scatter development	Built-up pixels that have neighbourhoods that are less than 30% built-up and not belonging to ribbon development

Table 6.3 Metrics for measuring attributes of sprawl (Angel et al. 2007)

Attribute	Metric	Definition
Urban extent	Built-up area	Impervious surface land-cover
	Urbanised area	Built-up area + urbanised open space
	Urbanised open space	Non-impervious pixels in which more than 50% of the neighbourhood is built-up
	Developable	Does not contain water or excessive slope
	Urban footprint	Built-up area + urbanised open space + peripheral open space
	Peripheral open space	Non-impervious pixels that are within 100 m of the built-up area
	Open space	The sum of the urbanised and peripheral open space
Density metrics	Built-up area density	City population divided by built-up area
	Urbanised area density	City population divided by urbanised area
	Urbanised area density restricted to developable area	City population divided by developable urbanised area
	Urban footprint density	City population divided by urban footprint
	Urban footprint density restricted to developable area	City population divided by developable urban footprint
Suburbanisation metrics	Cohesion	The ratio of the mean distance between a sample of points in the urbanised area and the mean distance among all points in a circle with an area equal to the urbanised area

6.5 Quantification and Characterisation of Sprawl   101

Table 6.3 (continued)

Attribute	Metric	Definition
	Decentralisation	The ratio of the mean distance to centre for all points in the urbanised area and the mean distance to centre for all points in a circle with an area equal to the urbanised area
	City centre shift	The distance between the original city centre (CBD) and the geometric urban centre (MAD centre)
	Minimum avg. distance (MAD) centre	The point with the minimum average distance to all other points in the urbanised area
	Density gradient	The exponent in the equation of the exponential trend-line fitting the points in the density versus distance from the CBD plot
Contiguity and openness metrics	New development	Built-up pixels existing in the land-cover for t_2 but not t_1
	New infill dev.	New development occurring within the t_1 urbanised open space
	New extension dev.	Non-infill new development intersecting the t_1 urban footprint
	New leapfrog dev.	New development not intersecting the t_1 urban footprint
	Openness index	The average percentage of open space within a 1 km^2 neighbourhood for all built-up pixels
	Open space contiguity	The probability that a built-up pixel will be adjacent to an open space pixel
	Open space fragmentation	The ratio of the combined urbanised and peripheral open space area to the built-up area
Compactness metrics	Single point compactness	The ratio of the area of the urbanised area and the area of the circle with the same average distance to the MAD centre of all pixels in the urbanised area

Table 6.3 (continued)

Attribute	Metric	Definition
	Constrained single point compactness	The ratio of the area of the urbanised area and the developable area in the circle with the same average distance to the MAD centre of all pixels in the urbanised area

Sierra Club (1998) ranked major metropolitans in USA by four metrics, including: population moving from inner city to suburbs; comparison of land-use and population growth; time cost on traffic; and decrease of open space. USA Today (2001) put forward the share of population beyond *standard metropolitan statistical area* (SMSA)[8] as an indicator for measuring sprawl. Smart Growth America (Ewing et al. 2002) carried out a research to study the impacts of sprawl on life quality in which four indices were used to measure urban sprawl: (1) residential density; (2) mixture of residence, employment and service facilities; (3) vitalization of inner city; and (4) accessibility of road network.

Some of the researchers also have contributed to measuring sprawl (Nelson 1999; Kline 2000; Torrens 2000; Galster et al. 2000; Glaeser et al. 2001; Hasse 2004) by establishing multi-indices by GIS analysis or descriptive statistical analysis. Those indices cover various aspects including population, employment, traffic, resources consumption, architecture aesthetics, and living quality, etc. Commonly used indices include: *growth rate*—such as growth rate of population or built-up area; *density*—such as population density, residential density, employment density; *spatial configuration*—such as fragmentation, accessibility, proximity; and *others*—such as per-capita consumption of land, land-use efficiency, etc. (e.g., The Brookings Institution 2002; USEPA 2001; Fulton et al. 2001; Masek et al. 2000; Pendall 1999; Sutton 2003; Jiang et al. 2007; Bhatta 2009a).

Many of the sprawl measurement are devised to reflect the relationship between population change and land conversion to urban uses. A hypothetical black-and-white sprawl determination approach is—if the built-up growth rate exceeds the population growth rate, there is an occurrence of sprawl (Sudhira et al. 2004; Bhatta 2009a). In a recent effort, the concept 'housing unit' was used as a proxy for population and combined with digital orthophoto data to generate urban sprawl metrics (Hasse and Lathrop 2003a). It is often difficult to distinguish population change in a given jurisdiction as either the cause or effect of urban development; therefore, 'the population factor should not be used as a sole indicator of urban sprawl' (Ji et al. 2006). In most cases, an increasing (or diminishing) number of built-up activities like housing and commercial constructions would be more effective to indicate

---

[8] A statistical standard developed for use by federal agencies in the production, analysis, and publication of data on metropolitan areas; each SMSA has one or more central counties containing the area's main population concentration and may also include outlying counties which have close economic and social relationships with the central counties.

sprawl as consequences of land consumption because usually construction activities, as compared to population change, reflect directly economic opportunities as the major driving force of land alteration (Lambin et al. 2001). Ji et al. (2006) have used landscape metrics to characterise sprawl using multi-temporal images. They have tried to formulate more effective metrics of urban sprawl by relating remotely sensed land change to construction-based land consumption. Bhatta (2009b) demonstrated a simple approach for black-and-white determination of sprawl—if the percentage of increase in growth rate of the city-extent exceeds the percentage of increase in built-up growth rate, it is an indication of sprawl.

Generally speaking, most of the aforementioned studies took whole city as the analysis unit to calculate the metrics, which could exactly reflect the sprawling situation of a whole city or region, but the interior differentia of sprawl in a given city could not be well depicted. Moreover, some indices are raised based on the context of developed countries; therefore, they are not so suitable for measuring sprawl in developing countries. Besides, some necessary statistical data are not successive enough to calculate certain indicator, such as the density of employment (Jiang et al. 2007).

The density of built-up area and intensity of annual growth could efficiently depict the sprawl features of low density and strong change, but they are still weak in capturing the particular spatial patterns of urban sprawl. Jiang et al. (2007) proposed 13 attributes (Table 6.4) under the name of 'geospatial indices' for measuring sprawl in Beijing. Finally they have proposed an *integrated urban sprawl index* that combines the preceding 13 indices. However, their approach requires extensive inputs of temporal data such as population, GDP, land-use maps, land-use master planning, floor-area ratio, maps of highways, and maps of city centres. Many developing countries lack such type of temporal data; and therefore, most of these indices are difficult to derive. Furthermore, they have not mentioned any threshold

Table 6.4 Definition of geospatial indices for measuring sprawl (Jiang et al. 2007)

Index	Description
Area index	Patch area of newly developed
Shape index	0.25 × perimeter/square root of area
Discontinuous development index	Distance between newly developed and previously developed land
Strip development index	Distance between newly developed patches and highways
Leapfrog development index	Distance between newly developed patches and county centres
Planning consistency index	{1, 0}, 1 stands for inconsistency with plan
Horizontal density index	Share of non-agricultural land area within neighbourhood of 1 km^2
Vertical density index	Ratio of floor area to land area
Population density index	Ratio of population to land area
GDP density index	Ratio of GDP to land area
Agriculture impact index	{1, 0}, 1 stands for arable land loss
Open space impact index	{1, 0}, 1 stands for open space loss
Traffic impact index	Population × distance to centres

to characterise a city as sprawling or non-sprawling. However, this type of temporal analysis is useful to compare between cities or different zones of a city or status of a city at different dates. Whether a city is becoming more sprawling or not, with the change of time, can be well depicted by this type of analysis.

Torrens (2008) argues that sprawl should be measured and analysed at multiple scales. In his (or her) approach to measuring sprawl, he has declared some ground-rules in developing the methodology. Measurements have been made to translate descriptive characteristics to quantitative form. The analysis is focused at micro-, meso-, and macro-scales and can operate over net and gross land. The analysis examines sprawl at city-scale and at intra-urban levels—at the level of the metropolitan area as well as locally, down to the level of land parcels. Although inter-urban comparison and use of remote sensing data are not focused on in this paper, the methodology should be sufficient to be generalised to other cities using remote sensing data. The methodology devised a series of 42 measures of sprawl, which are tracked longitudinally across a 10-year period. Although the author claims that this approach can provide a real insight of urban sprawl, however, the use of many scales has made the methodology more complex which may result in confusion (Bhatta et al. 2010b).

Entropy method, another urban sprawl metric, is perhaps the most widely used technique to measure the extent of urban sprawl with the integration of remote sensing and GIS (Yeh and Li 2001a; Lata et al. 2001; Li and Yeh 2004; Sudhira et al. 2004; Kumar et al. 2007; Bhatta 2009a; Bhatta et al. 2010a). Since it is the most preferred technique, let us discuss this measure in a greater detail.

Shannon's entropy (H_n) can be used to measure the degree of spatial concentration or dispersion of a geographical variable (x_i) among zones (Theil 1967; Thomas 1981). Entropy is calculated by:

$$H_n = -\sum_{i=1}^{n} P_i \log_e (P_i) \qquad (6.1)$$

where P_i is the probability or proportion of a phenomenon (variable) occurring in the ith zone ($P_i = x_i / \sum_{i=1}^{n} x_i$), x_i is the observed value of the phenomenon in the ith zone, and n is the total number of zones. The value of entropy ranges from 0 to $\log_e(n)$. A value of 0 indicates that the distribution of built-up areas is very compact, while values closer to $\log_e(n)$ reveal that the distribution of built-up areas is dispersed. Higher values of entropy indicate the occurrence of sprawl. Half-way mark of $\log_e(n)$ is generally considered as threshold. If the entropy value crosses this threshold the city is considered as sprawling.

Relative entropy can be used to scale the entropy value into a value that ranges from 0 to 1. Relative entropy (H'_n) can be calculated as (Thomas 1981):

$$H'_n = \frac{H_n}{\log_e (n)} \qquad (6.2)$$

In this instance 0.5 is considered as threshold.

Because entropy can be used to measure the distribution of a geographical phenomenon, the measurement of the difference of entropy between time t_1 and t_2 can be used to indicate the magnitude of change of urban sprawl (Yeh and Li 2001a):

$$\Delta H_n = H_n(t_2) - H_n(t_1) \qquad (6.3)$$

The change in entropy can be used to identify whether land development is toward a more dispersed (sprawled) or compact pattern.

Yeh and Li (2001a) have shown how to use the entropy method to measure rapid urban sprawl in fastest growing regions. Bhatta et al. (2010a) proposed an approach that shows how the entropy method can be used to measure the sprawl as a pattern, process, and 'overall' that combines both pattern and process. They have calculated the entropy from a built-up growth matrix for the rows, columns, and then both. However, the sprawl phenomenon relates to the dispersion of built-up area, therefore, the use of built-up growth (instead of built-up area) to determine the entropy is a questionable approach (Bhatta et al. 2010b). The basic approach of Yeh and Li (2001a) is rather more acceptable.

Entropy is a more robust spatial statistic than the others (Yeh and Li 2001a). Many studies have shown that spatial dispersal statistics, such as the *Gini* and *Moran coefficients* (Tsai 2005), which are often dependent on the size, shape, and number of regions used in calculating the statistics; and the results can change substantially with different levels of areal aggregation (Smith 1975). This is a manifestation of the scale problem or *modifiable areal unit problem* (refer Chap. 8) which may exert unspecified influence on the results of spatial analysis (Openshaw 1991). Thomas (1981) indicates that relative entropy is better than traditional spatial dispersal statistics because its value is invariant with the value of n (number of regions). However, relative entropy is still, to some extent, sensitive to the variations in the shapes and sizes of the regions used for collecting the observed proportions. For example, if there are two scales of analysis for the dispersion of population in a country, such as regions and sub-regions, different entropy values will be obtained if the data are collected based on regions instead of sub-regions. The entropy decomposition theorem can be used to identify different components of the entropy that are related to different zone sizes in collecting the data (Batty 1976; Thomas 1981; Yeh and Li 2001a). It can alleviate the problem for comparing the results between different zone sizes because the influence of scaling can exactly be measured (Yeh and Li 2001a).

Some other urban growth/sprawl studies:
Bhatta (2009a) shows how the built-up data derived from remote sensing imageries can be used to analyse and model the urban growth, such as, proportion of population and proportion of built-up area; growth rate of population and built-up areas; proportion of household and proportion of built-up areas;

urban growth, existing built-up area, developable land, and change in number of working persons; and entropy method to measure the sprawl.

Bhatta et al. (2010a) shows how to calculate the expected growth from the observed urban growth using built-up information derived from remote sensing data. This study also shows how to calculate the degree-of-sprawl, degree-of-freedom and degree-of-goodness of urban growth. All of these measures have been applied in three different dimensions—pattern, process, and 'overall'. Degree-of-freedom reveals the freedom or degree of deviation for the observed urban growth over the expected. Higher degree-of-freedom for a zone is an indication of unstable development within the zone with the change of time. And higher degree-of-freedom for a temporal span can be considered as higher inter-zone variability in urban growth. The degree-of-goodness actually refers to the degree at which observed growth relates the expected growth and the magnitude of compactness (as opposed to sprawl).

Davis and Schaub (2005) have applied impervious surface metric, neighbourhood density metric, and building permit metric for quantifying urban growth and sprawl from two temporal remote sensing imageries.

Irwina and Bockstael (2004) have used parcel data on residential land conversion to investigate how land-use externalities influence the rate of development, and modify policies designed to manage urban growth and preserve open space. Several smart growth policies are found to significantly influence land conversion, including a development clustering policy that concentrates development and generates preserved open space. Although they have not used remote sensing imageries, however, their approaches are similarly applicable on remote sensing data if parcels can be mapped from high resolution remote sensing imageries.

Wilson et al. (2003) have used land-cover maps, derived from remotely sensed satellite imagery, to determine the geographic extent, patterns, and classes of urban growth over time. The model identifies three classes of undeveloped land as well as developed land for both dates based on neighbourhood information. These two images are used to create a change map that provides more detail than a traditional change analysis by utilising the classes of non-developed land and including contextual information. The change map becomes the input for the urban growth analysis where five classes of growth are identified: infill, expansion, isolated, linear branch, and clustered branch.

Sutton (2003) has used nighttime satellite imagery to measure the areal extent of the urban areas. The nighttime satellite imagery was compared to a gridded population density dataset derived from the census data. This comparison resulted in measures of both the areal extent and the population of all the urban areas. These numbers were then used to calculate improved 'scale-adjusted' aggregate measures of urban sprawl. A similar study can be found in Imhoff et al. (1997).

Chapter 7
Modelling and Simulation of Urban Growth

7.1 Introduction

A *model* is a simplified representation of the physical system. They are the tools to simulate the behaviour of physical systems. They can predict the future evolution of the systems, they can be used as interpretative tools in order to study system dynamics, and they can give hints for data collection and design of experiments (Giudici 2002).

A simplified definition of model is—theoretical abstractions that represent systems in such a way that essential features crucial to the theory and its application are identified and highlighted. Models are basically built by consideration of the pertinent physical principles operated on by logic and modified by experimental judgment and plain intuition. In this role, models act as a vehicle to enable experimentation with theory in a predictive sense, and to enhance understanding which may be prior to predictions of situations.

Urban modelling is the activity of defining, building, and applying models for specific purposes which traditionally have been in physical planning. These applications increasingly extend to other social and human geographies built around location theory and spatial analysis. The measurement of urban change is an obvious extension of urban change detection. Contemporary research also includes the use of models to simulate the urban growth and predict future urban dynamics. These models are called *urban growth models*.

This chapter is aimed to discuss on urban models; in particular, it identifies distinct classes of model, beginning with theoretical models to cell-based, agent-based, and rule-based land-use transport models; their merits and demerits; and finally the chapter ends with a detailed discussion on studies for modelling of urban growth from remote sensing data.

7.2 Urban Model and Modelling

Urban models are representations of functions and processes which generate urban spatial structure in terms of land-use, population, employment and transportation, usually embodied in computer programs that enable location theories to be tested against data and predictions of future locational patterns to be generated (Batty 2009). This suggests that urban models are essentially computer-based simulations used for testing theories about spatial location and interaction between land-uses and related activities. They also provide digital environments for testing the consequences of physical planning policies on the future form of cities. The main objectives of model are: (1) to integrate observations, information and theories concerning a system; to aid understanding of system behaviour; (2) to predict the response of the system to the future changes; and (3) to allocate certain resources in order to optimise certain conditions within the system.

Urban modelling is the process of identifying appropriate theory, translating this theory into a mathematical or formal model, developing relevant computer programs and then confronting the model with data so that it might be calibrated, validated and verified prior to its use in prediction (Batty 2009). Urban systems and their phenomena are very complex; therefore, it is practically impossible to model such systems perfectly, and any model of these systems requires simplification and approximation. Figure 7.1 represents a simplified conceptual schema of process of model

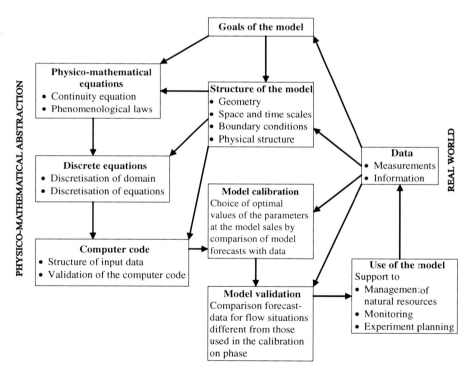

Fig. 7.1 Schema of the process of development and application of a physically based model (Giudici 2002)

development and its application. However, important to realise that reality is far more complex, and model development and application is not a straight forward process.

7.2.1 Classification of Urban Models

Urban modelling has a long history (refer Batty 2008). Over the last 50 years since computer models were first developed in the urban domain (Harris 1965), several distinct types have been emerged, which can be classified from different perspectives (Batty 2009; Pooyandeh et al. 2007): Static versus dynamic; macro and micro; cell-based and zone-based; mathematical implementation (linear and nonlinear); complexity and temporal; deterministic or probabilistic; economic or social or mixed; etc. Despite several classification schemes as stated, Batty (2009) has suggested to group them into three main classes: (1) land-use-transportation model; (2) urban dynamics model; and (3) cellular automata, agent-based model and micro-simulation. However, this classification also has several problems; for example, cellular automata and agent-based models are different. Further, it does not consider neural networks and fractal based modelling.

Pooyandeh et al. (2007) have classified spatio-temporal urban models into two distinct classes: (1) complexity model, and (2) temporal GIS model. Complexity model is further sub-divided into: CA based modelling, agent based modelling, neural network based modelling and fractal based modelling. However, they have also overlooked several other models, for example, theoretical models.

The review of literature has found that there are several other novel models that are of distinct type or of mixed in nature. Therefore, rather than following any of the preceding classification scheme, it is better to discuss these models as follows.

7.3 Theoretical Models

A scientific theory can be thought of as a model of reality or physical system. Theoretical models are based on mathematics and physics; and attempt to make general theories about urban processes. These models are used to test and establish theoretical hypothesis. They can often explain processes in just one city, and only during a certain period of time. Schweitzer has done work using the concepts of Brownian motion to develop theoretical models of self organisation within urban and human processes (Schweitzer and Holyst 2000; Schweitzer and Tilch 2002; Schweitzer 2002, 2003). Since these models are based on theory, they cannot provide useful information to planners and decision makers. These models have little or no use beyond the limit of research frontier.

7.4 Aggregate-Level Urban Dynamics Models

Urban dynamics are representations of changes in urban spatial structure through time which embody a myriad of processes at work in cities on different but often interlocking time scales ranging from life cycle effects in buildings and populations to movements over space and time as reflected in spatial interactions (Batty 2009). Worth mentioning that in this event urban dynamics model refers to aggregate-level urban dynamics model—i.e., macro level (Forrester 1969; Allen 1997; Wilson 2000; Dendrinos and Mullally 1985). Very few aggregate-level urban dynamics models have been applied empirically. They were simply the path to more sophisticated micro-level models and have been eclipsed somewhat by spatial simulations of dynamic processes whose scale is at a much more individualistic level as embodied in agent-based modelling (Batty 2009).

7.5 Complexity Science-Based Models

Complexity science studies the common properties of systems that are considered fundamentally complex. Such systems are used to model processes in biology, economics, physics and many other fields. Complex Systems is a new approach to science that studies how relationships between parts give rise to the collective behaviours of a system and how the system interacts and forms relationships with its environment.

Due to a number of characteristics, urban systems are also considered as complex systems[1] (Alexander 1965; Jacobs 1965; McHarg 1969). These distinguishing qualities can be named as *fractal dimensionality, self-similarity, self-organisation* and *emergence* (Armstrong 1988; Batty and Longley 1994; Portugali 2000; Shen 2002; Torrens 2000). For addressing the complex problems in city planning, it is not sufficient just to be concerned with the physical structure of the city; rather the interplay of intangible economic, social and environmental factors needs to be considered as well (Gilbert et al. 1996). Therefore linear and reductionist science is of limited value in urban modelling. This trend grew to produce the ideas of complexity science based models.

The main reason for adopting complexity models is that the traditional urban models do not address some important features required for urban applications (Pooyandeh et al. 2007). Torrens (2001) identified following weaknesses for traditional urban models: their centralised approach, a poor treatment of dynamics, weak attention to detail, shortcomings in usability, reduced flexibility, and a lack of realism. These models are incapable to address the concerns of current planning

---

[1] Systems that show surprising and unanticipated or 'emergent' behaviours as shown in patterns that arise at the aggregate level from the operation of system processes at the micro or agent level. Such systems are intrinsically unpredictable in an overall sense but can be fashioned in such a way that makes knowledge of them useful and certain. Cities are the archetypical example, but so too is the economy. (Batty 2009).

and policy analysis, which are the issues like regeneration, segregation, polarisation, economic development, and environmental quality (Batty 2003). In the following sections these models are discussed according to their applications; and merits/demerits of each one are presented.

7.5.1 Cell-Based Dynamics Models

Cell-based dynamics models are developed from the ideas of complexity science based on *cellular automaton* (CA) (plural: *cellular automata*) (Ilachinski 2001). CA is a class of spatially disaggregate models, often pictured as being formed on a two-dimensional lattice of cells, where each cell represents a land-use (or land-cover) and where embodying processes of change in the cellular state are determined in the local neighbourhood of any and every cell (Batty 2009). Traditional form of CA consists of five main components: the lattice, time steps, cells, a neighbourhood, and transition rules. CA is formally defined as follows (Pooyandeh et al. 2007; Batty 2009):

$$S^{t+1} = f(S^t, S_N) \tag{7.1}$$

where S^{t+1} is the state of the cell at time $t+1$, S_N is the set of states of neighbourhood, f is a function representing a set of transition rules, and S^t is the state of the cell at time t. Time is divided into discrete time steps, and the transition rules examine the cells in the neighbourhood to determine if and how the current cell will change state at the next time step.

These models are able to handle large amounts of data and many fields of study such as population, economic activity and employment, land-use and transport (Batty 2005). They can deal with diverse choices of priorities, producing scenarios covering many impacts. These models indicate the range of possible impacts. Also, depending on the rules, they can cover issues like environmental impacts of economic decisions. They can be tailored to deal with a specific city. Diappi et al. (2004) have identified four distinct advantages of CA: (1) spatial inherency—definition on a raster of cells and on neighbouring relationships are crucial; (2) simple and computationally efficient; (3) process-based modelling that deal with state changes; and (4) dynamic in nature and can represent a wide range of situations and processes. Al-Ahmadi et al. (2009) also list several advantages of CA.

Reviewing the applications of CA modelling in urban domain, clarifies the degree of suitability of this technique in urban applications. For example, diffusion or migration of resident populations (Portugali et al. 1997), competitive location of economic activities (Benati 1997), joint expansion of urban surface and traffic network (Batty and Xie 1997), generic urban growth (Clarke et al. 1997), urban evolution simulation (Batty 1998; Wu and Martin 2002), self-organising competitive location simulation (Benati 1997), polycentricity of urban growth (Wu 1998), urban simulation (Torrens and O'Sullivan 2001), simulation of urban dynamics (Itami 1994), emergent urban form (Xie and Batty 1997), land-use dynamics (Deadman et al. 1993; Batty and Xie 1994; Phipps and Langlois 1997; White and Engelen 1997; White et al. 1998; White and Engelen 1993; Cecchini 1996; Batty et al. 1999), real estate investment simulation (Wu 1999), long term urban growth prediction (Clarke and Gaydos 1998), regional scale urbanisation (Semboloni 1997; White and Engelen 1997), urbanism (Sanders et al. 1997), simulation of development density for urban planning (Yeh and Li 2002), urban socio-spatial segregation (Portugali 2000), urban development plan (Chen et al. 2002), landscape dynamic (Soares-Filho et al. 2002), modelling of urban development (Wu and Webster 1998; Liu and Phinn 2003), and self-modifying CA to model historical urbanisation (Clarke and Hoppen 1997), are some of its numerous applications. The popularity of CA for urban applications is due to the fact that they are much more straightforward than the differential equations (Portugali 2000), and they are compatible with GIS systems and remotely sensed data (White and Engelen 2000), also, connecting the form with function, and pattern with process they are flexible and dynamic in nature. Such models can be seen as simplifications of *agent-based models* where the focus is on emergent spatial patterns through time.

Despite the growing tendency towards using CA in urban modelling, the degree of fitness of these models in some urban applications is questionable (refer Yeh and Li 2003; Pooyandeh et al. 2007). One may refer Clarke and Gaydos (1998) that documents several modifications on CA which have been proposed to overcome the problems involved in application of CA in urban fields. One of the problems mentioned is that despite the similarities between urban systems and living organisms they have some basic differences. In fact, many stakeholders contribute to the change of state in an urban structure, so the combination of their activities causes 'state transitions' (O'Sullivan and Torrens 2000). Another problem is regular shapes of cell which do not match with the basic units of urban systems like land parcels. To solve the problem, there have been a number of efforts in using cells with irregular shapes (O'Sullivan 2002; Pang and Shi 2002). However, irregular shaped cells are not compatible with remote sensing data.

The presentation of time in CA is also problematic because synchronous update of cell states does not match with many urban systems. To overcome this deficiency, asynchronous update has been proposed (Wu 1999). Something that should be added is that in urban applications the impacts of basic units on each other is not limited to the neighbourhoods, but the classic CA concept does not allow 'action-at-a-distance' (O'Sullivan and Torrens 2000). There have been several experiments with the use of a variety of neighbourhood sizes and types and different cell dimensions to overcome the problem (Jenerette and Wu 2001; White and Engelen 2000). The main focus of this book is of course on urban growth which in a contemporary manifestation is sprawl with these models CA tending to be indicative rather than predictive (Batty 2009). The other issue in such models is that they are primarily physicalist[2] in scope and as such, largely ignore features of the spatial economy

---

[2] This concept assumes that everything which exists is no more extensive than its physical properties; this concept means that there are no kinds of things other than physical things.

such as house prices, wage rates, and transport costs. Contribution of diverse agents in an urban environment proposed the idea that using agent-based modelling in some urban applications seems more appropriate.

7.5.2 Agent-Based Models

Agent-based models are a class of models developed since the 1980s which are based on representing objects and populations at an elemental or individualistic level which reflect behaviours of those objects through space and time (Batty 2009). Unlike CA models that operate in a top-down way, agent-based models operate in a bottom-up way and sometimes generate emergent spatial and temporal patterns at more aggregate levels. These models can simulate the simultaneous operations of multiple agents, in an attempt to re-create and predict the actions of complex phenomena. The process is one of emergence from the lower (micro) level of systems to a higher (macro) level.

Agent-based models are also based on complexity science, however, rather than focusing on cells (as in CA) they are based on the emergent properties of the actions of mobile agents that move through the city, which represent major players in a city's development. *Agents* are identifiable, mobile, autonomous, flexible and goal-directed individuals, which interact with each other and their behaviours (refer Pooyandeh et al. 2007), are managed by predefined rules. Usually the actions of agents, people, businesses, etc., are aggregated to simplify the calculations. The main advantage of this style of urban growth modelling is that it is dynamic and behavioural and can easily extend to both demand and supply sides of the development process. Castle and Crooks (2006) identified several advantages for agent-based models.

Studies on agent-based models include: simulation of residential dynamics in the city (Benenson 1998), dynamics of pedestrian behaviour in streets (Batty 2001; Kerridge et al. 2001; Schelhorn et al. 1999), modelling the discrete dynamics of spatial events for mobility in carnivals and street parades (Batty et al. 2003), integration of CA and agent-based models to support the exploration of 'what-if' scenarios for urban planning and management (Torrens 2001), and interactive multi-agent simulation model on the web ('CityDev' of Semboloni et al. 2004). Other examples of agent based models are Portugali (2000) and Torrens (2003). Najlis and North (2004) discuss that there is an increasing interest in the integration of GIS and agent-based modelling systems (e.g., Brown et al. 2005; Parker 2004; Torrens and Benenson 2005).

Agent-based models have their uses as mentioned in the preceding paragraph, but are limited in their ability to deal with land-use changes. Furthermore, to fully model a complex system, many attributes and behaviours should be considered for an agent. It has enormous data requirements, and does not fit easily into more top-down processes that drive the urban system. Because heterogeneity in agents is often introduced by local randomness, it does not generate the sorts of deterministic prediction that are usually needed in operational urban modelling (Batty 2009). To overcome the problem of extensive data requirements multiple runs and varying initial conditions are required (Axtell 2000), this brings about computational problems and makes the system more complex. On the other hand they are very sensitive to initial conditions and to small variations in interaction rules (Couclelis 2002). Finally, although agents have made outstanding improvements in modelling systems that demonstrate subjective and unpredictable behaviours like that of human beings it is clear that a lot should be done for a satisfactory simulation of such systems (Pooyandeh et al. 2007).

> If one adds individual actors to CA environment, then one has urban models which are agent-based. This may confuse the discrimination between them. In many instances, CA models are paraphrased as agent-based models. Agent-based models have all the characteristics of CA models; in addition to that, they can also be programmed for spatial mobility and to represent the external factors responsible for the processes. They differ in terms of their mobility: CA cannot 'move', but agent-based models are mobile entities. They also differ in terms of interaction, CA transmit information by diffusion over neighbourhoods; whereas agent-based models transmit information by themselves, moving between locations that can be at any distance from an agent's current position (Torrens 2006). Unlike the CA models, behaviour of agent-based models can be scheduled to take place synchronously or asynchronously (Castle and Crooks 2006). So in CA, cells are agents and vice versa, but in agent-based modelling agents and cells are totally different (Batty 2003).

7.5.3 Artificial Neural Network (ANN)-Based Models

With the development of ANN, which are *Artificial Intelligence* based information processing systems, in recent years new opportunities have emerged to enhance the tools we use to process spatial data. ANN models try to imitate the structure of human brain through mathematics. An ANN model is a mathematical model or computational model that tries to simulate the structure and/or functional aspects of biological neural networks. It consists of an interconnected group of artificial neurons and processes information using a connectionist approach to computation. In most cases an ANN is an *adaptive system* that changes its structure based on external or internal information that flows through the network during the learning phase. ANN models are non-linear statistical data modelling tools. They can be used to model complex relationships between inputs and outputs or to find patterns in data.

7.5 Complexity Science-Based Models

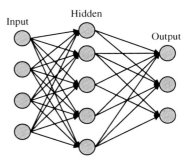

Fig. 7.2 ANN is an interconnected group of nodes consisting three layers

As described by Pooyandeh et al. (2007), ANN has a number of *nodes* which are called *processing elements* or *processing units*. A state level and a real-valued bias are associated with each node. The edges of the network are called *connection* that joins two nodes. A weight is associated with each connection which determines the strength of this link. The network consists of *input*, *output* and *hidden* layers (Fig. 7.2). Numeric data moves from each unit to another through the connections. The node is broken into two components, the *internal activation function* and the *transfer function* (DuBose and Klimasauskas 1989). Internal activation typically operates through a summation type function which adds up the values of incoming messages multiplied by a specific connection weight. The resultant output of the internal activation is sent to the transfer function. The transfer function, which is a linear or non-linear algebraic function, is defined for each node that determines the state of that node based on its bias, the weights on its incoming links from other nodes, and the different states of the nodes that are connected to it via these links (Openshaw 2000).

A number of ANN-based applications (e.g., Bishop 1994; Gimblett and Ball 1995; Abrahart and Kneale 1997; Rigol et al. 2001; Fischer et al. 1997; Rodrigue 1997; Shmueli 1998; Kropp 1998; Pijanowski et al. 2002) can be found in the existing literature (refer Pooyandeh et al. 2007; Diappi et al. 2004; Openshaw 2000). Liu (2000) has applied ANN to detect the change from non-urban to urban land-use. Diappi et al. (2004) have used ANN for prediction and for building virtual scenarios on urbanisation processes. Li and Yeh (2001) used ANN to determine the CA simulation parameters. They imported the parameter values from the training of ANN into the CA models. Also other studies were done by Yeh and Li (2003, 2004) to apply ANN for determining CA parameters.

Fischer (1995), Diappi et al. (2004), and others have identified several advantages of ANN; for example, (1) parallel information processing of ANN speeds up processing noticeably; (2) ANNs are capable of decision making based on noisy and fuzzy information; (3) knowledge is encoded in patterns of numerical strength associated with the connections that exist between the processing elements of the system (Smolensky 1998); and (4) non-linear structure enables prediction and performing 'what-if' scenarios (Corne et al. 1999). However, using of ANN models is not so straightforward and a relatively good understanding of the underlying theory is essential. Complex ANN models tend to lead to problems with learning (Liu and Lathrop 2002). There are numerous tradeoffs between learning algorithms. Almost any algorithm will work well with the correct parameters for training on a particular fixed dataset. However selecting and tuning an algorithm requires a significant amount of experimentation.

7.5.4 Fractal Geometry-Based Models

Fractal geometry is a rough or fragmented geometric shape that can be split into parts, each of which is (at least approximately) a reduced-size copy of the whole (Mandelbrot 1982). A *Fractal* is defined as 'a set for which the Hausdorff-Besicovitch dimension strictly exceeds the topological dimension' (Mandelbrot 1983). Applications of fractals in urban modelling are based on the notion that some designed spatial objects show fractal qualities. Some recent fractal-based urban modelling are Batty (1994), Shen (2002), White and Engelen (1993), Yeh and Li (2001b), and Frankhauser (1994). Refer Pooyandeh et al. (2007) for more detailed discussion. Important to mention that fractal geometry-based models are not concerned with processes, but only with application of fractal patterns to city modelling. As these models do not consider micro processes in cities, they have a limited use for decision making. Further, the blindness to process has made these models unusable for urban growth.

7.6 Rule-Based Land-Use and Transport Models

Rule-Based Land-use and transport models focus primarily on the way populations and employments locate in urban space consistent with the spatial interactions between different locations of these activities (Batty 2009). These models usually simulate the city at a cross-section in time (static) and do not consider urban dynamics (Fujita et al. 1999; Ben Akiva and Lerman 1985; Waddell 2002). Furthermore, these models are based on a top-down approach; they fail to represent the complexity of cities, the variations within a city and the dynamic nature of change. Since these models do not embrace urban dynamics, they cannot be applied effectively in urban growth modelling and they are generally used in urban transportation modelling.

CUF (California Urban Futures) model is a CA based model for changing land-use (Landis and Zhang 1998).

'What If?' is a model that links with GIS to examine urban change scenarios, especially in cities experiencing rapid growth (Klosterman 1999).

7.6 Rule-Based Land-Use and Transport Models

QUEST is a rule-based model, developed by the Sustainable Development Research Initiative in Vancouver to explore sustainable futures. QUEST allows people to develop 'what if?' scenarios for the future of the region. Its purpose is to aid the public and policy makers to explore options for a sustainable future. It handles a large amount of data about the region and works through deterministic formulae based on best knowledge at land-use and transport use and their environmental impacts (Waddell and Evans 2002).

UrbanSim is a rule-based land-use transport model looking at land-use and transport-use and their environmental impacts (Waddell and Evans 2002). UrbanSim class of models is now being fashioned into the Open Platform for Urban Simulation (OPUS) which is located at http://www.urbansim.org/. Refer Noth et al. (2003) for a detailed discussion on UrbanSim model.

DUEM (Dynamic Urban Evolutionary Model) is a CA based model developed in London and Michigan, which can be used to develop simple demonstrations of cellular growth (available at http://www.casa.ucl.ac.uk/software/duem.asp).

SLEUTH (Slope, Land-use, Exclusion, Urban extent, Transportation and Hillshade) is another CA based popular model developed by Keith Clarke (Clarke et al. 1997; Silva and Clarke 2002) under the United States Geological Survey for simulating urban growth in North America (available at http://www.ncgia.ucsb.edu/projects/gig/).

CAST (City Analysis Simulation Tool), another model based on complexity and CA, is a non-predicative tool which gives a range of future scenarios of possibilities in 10 or 15 years exploring different options of what urban development and change may look like (available at http://www.intesys.co.uk/cast/index1.htm).

SIMLUCIA is another CA based model for estimation of area for land-use transition conceived by White et al. (1998). This is an integrated model of natural and human systems operating at several spatial scales, and was aimed at providing the officials of the Caribbean Island of Santa Lucia with a tool to explore possible environmental, social and economic consequences of hypothesised climate change (available at http://www.spatial-modelling.info/SimLucia-climate-change-and).

DINAMICA is a CA-based simulation model developed by the Centre for Remote Sensing of the Federal University of Minas Gerais (Almeida et al. 2005). This model can simulate urban land-use change from remote sensing data.

UES (Urban Expansion Scenario), a CA-based model proposed by He et al. (2006), is an urban growth simulation tool from remote sensing data.

FCAUGM (Fuzzy Cellular Automata Urban Growth Model) is an urban growth simulation model that uses remote sensing data, CA, and fuzzy set theory (Al-Ahmadi et al. 2009, 2008a, 200b).

> LUCAS is model to assess the impact of urban expansion on the surrounding natural areas by using landscape metrics (Berry et al. 1996). Generally, this approach can be applied in a variety of investigations relating to urban dynamics and the resulting spatial structures.

7.7 Modelling of Urban Growth

Socioeconomic, natural, and technological processes both drive and are affected by the evolving urban spatial structures within which they operate. In the recent years, models of land-use/land-cover change and urban growth have become important tools for city planners, economists, ecologists and resource managers (Agarwal et al. 2000; EPA 2000; Klosterman 1999; Wegener 1994). This development has been possible owing to an increased availability and usability of multiple spatial datasets, development of computer hardware and software tools for processing a wide variety of data. Community-based collaborative planning and consensus-building efforts in urban development have also been strengthened by the new data and tools at the local level (Klosterman 1999; Sui 1998; Wegener 1994). There are two basic reasons for urban growth modelling: first is the need to increase the level of understanding of the cause-effect relationships operative in urban growth dynamics, and secondly, to apply that increased understanding to aid the decision making process for the urban growth management.

7.7.1 Modelling of Urban Growth from Remote Sensing Data

Although urban modelling is not a new concept and has long history (Batty 2008); modelling of urban growth and sprawl has not been widely practiced, especially in developing countries. Several problems have been identified in building, calibrating and applying models of urban growth and urban land-use/land-cover change. These relate to the issues of data availability and to the need for improved methods and theory in modelling urban dynamics (Irwin and Geoghegan 2001; Longley and Mesev 2000; Wegener 1994). In general, the quality of results obtained from models is strongly depends on the quality and scope of the input data, calibration[3] and validation[4] (Batty and Howes 2001; Longley and Mesev 2000; Giudici 2002).

---

[3] Model calibration is to determine the numerical values of model parameters, given some data set. A calibrated model is often believed to be a tool ready to forecast a system's behaviour.

[4] Model validation assesses the model's ability to predict the behaviour of the system under conditions different from those used in the calibration phase. This is usually done by comparing the model forecasts and the observations for a time period following that for which the model was originally calibrated.

7.7 Modelling of Urban Growth

Since many of the urban models simulate both human and environmental systems, the requirements placed on the data are fairly complex and range from natural and ecological to socioeconomic data and large-scale land-use/land-cover data with adequate spatial and temporal accuracy. Socioeconomic data sources include census and various other types of governmental data as well as data that are routinely collected by local planning and administrative agencies (Fagan et al. 2001; Foresman et al. 1997; Wegener 1994).

However, 'these data sources are generally limited in their temporal accuracy and consistency, in their inclusion of important urban variables, and in their availability for different areas, especially outside the developed countries' (Herold et al. 2005a). Accordingly, a number of studies have explored alternative sources of data, in specific data from remote sensing sensors (Acevedo et al. 1996; Clarke et al. 2002; Meaille and Wald 1990). These studies found that remote-sensing techniques can provide reliable and spatially consistent datasets over large areas with both high spatial and temporal resolutions, including historical archives. Remotely sensed data can represent urban characteristics and structure such as spatial extent and pattern of land-cover, sometimes inferences about land-use and urban infrastructure, and indirectly, a variety of socioeconomic patterns (refer Usher 2000). Spatial metrics and spatial statistics have been used to evaluate and assess the local, small-scale performance of models in addition to the summary statistics addressing total amounts of change or growth. Spatial metrics assess the goodness of fit in terms of spatial structure and highlight specific problems, uncertainties or limitations of the model results (Clarke et al. 1996; Herold et al. 2003a; Manson 2000; Messina et al. 2000).

Our understanding of physical and socioeconomic patterns and processes through urban modelling is limited by the available data. Longley and Mesev (2000) refer to remote sensing as an important and insufficiently exploited source of data to aid not only urban applications but also theoretical understanding. Batty and Howes (2001) also argue that remote sensing data can provide a unique insight on spatial and temporal change of urban patterns. Remote sensing may also contribute to better representations of the spatial heterogeneity of urban land-use structure, landscape features and socioeconomic phenomena, improving on the traditional models that often tend to reduce urban space to a unidimensional measure of distance (Irwin and Geoghegan 2001). However, the potentials of the combined application of remote sensing techniques and urban modelling have yet to be fully explored and evaluated (Batty and Howes 2001; Longley and Mesev 2000; Longley et al. 2001; Herold et al. 2005a).

The most frequently used model to simulate the urban growth is CA (Torrens 2000). Perhaps the most popular CA urban growth model is SLEUTH (Clarke et al. 1997). SLEUTH can be used to model historical growth (Clarke and Gaydos 1998) as well as to predict future urban extent and pattern. This prediction is based on 'what if' scenarios (Clarke et al. 1997). Candau and Goldstein (2002) have used this model for investigating five growth options for Santa Barbara, California, to model change over three decades. They have used spatial metrics to quantify and describe the patterns produced under the various management plots. Yang and Lo (2003) have shown the use of SLEUTH to model the future urban growth of Atlanta,

Georgia. Other studies based on SLEUTH include Silva and Clarke (2002), Yang (2002), Yang and Lo (2002), Syphard et al. (2005), Jantz et al. (2003), Xian and Crane (2005), Caglioni et al. (2006).

Yang et al. (2008) tested the support vector machines (SVM) as a method for constructing nonlinear transition rules for CA based on remote sensing and GIS data. The proposed SVM-CA model was used for simulating the urban development in the Shenzhen City, China. They claimed that the SVM-CA model can achieve high accuracy and overcome some limitations of existing CA models in simulating complex urban systems. Almeida et al. (2005) have used CA-based DINAMICA model to simulate the urban land-use change from remote sensing data for a medium-sized town in the west of São Paulo State, Bauru. Different simulation outputs for the case study town in the period of 1979–1988 were generated by this empirically devised model; statistical validation tests were then conducted for the best results, employing a multiple resolution fitting procedure.

Alkheder and Shan (2005) also demonstrated a CA-based model to simulate the urban growth for Indianapolis city, Indiana. They have used historical remote sensing data to simulate the future. These simulated future maps were then compared with the respective future satellite images to validate the model. They report average accuracy of the simulated results is over 80% (with average 85% classification accuracy). He et al. (2006) introduced another CA-based urban growth simulation model that can operate on remote sensing data. They have applied this model for the city of Beijing. They have also coupled this CA-based model with a system dynamic model that operates in a top-down way to predict future urban growth. The system dynamic model was based on 'what-if' scenario. They have predicted the future growth of Beijing for different years up to 2020. Although they claim that the system dynamic model can be an appropriate approach to reflect the driving forces and analyse the implications of different policies; however, ability of this model to represent the spatial process of land-use is weak because it can not deal with spatial data well and can not effectively describe the detailed distribution and situations of the spatial factors in the land system (Guo et al. 2001; Zhang 1997).

Al-Ahmadi et al. (2009, 2008a, b) presented FCAUGM model for the simulation of urban growth. This model is based on CA; however, it uses fuzzy logic to handle the complexity of urban system. This model was demonstrated to simulate the urban growth of Riyadh, Saudi Arabia by using remote sensing data. They claim that instead of using probability theory, as in previously reported CA urban models, fuzzy logic can allow better understanding of complex social, physical and economic factors which influence urban growth.

Yeh and Li (2004) presented a new approach of integrating CA and ANN for simulating alternative land development of urban growth using remote sensing data. Alternative development patterns can be formed by using different sets of parameter values in CA simulation. They have demonstrated that neural networks can simplify CA models but generate more plausible results. The simulation is based on a simple three-layer network with an output neuron to generate conversion probability. No transition rules are required for the simulation. Parameter values are automatically obtained from the training of network by using remote sensing data.

7.7 Modelling of Urban Growth

Original training data can be assessed and modified according to planning objectives. Alternative urban patterns can be easily formulated by using the modified training data sets rather than changing the model.

Cheng and Masser (2003) presented a spatial data analysis method to seek and model major determinants of urban growth in the period 1993–2000 by a case study of Wuhan City in China. This method, based on remote sensing data, comprises exploratory data analysis and spatial logistic regression technique.

Allen and Lu (2003) took a hybrid approach to modeling and prediction of urban growth in the metropolitan Charleston region. To make the prediction objective flexible and realistic, a binary logistic model was integrated with a rule-based suitability model through a participatory process to predict the probabilities of urban transformation. Future urban growth was simulated based on different growth scenarios. The researchers have justified the use of a multiple-model approach for long-term prediction.

Finally, the results of spatial modelling need to be thoroughly interpreted and assessed in order to derive useful information for specific applications. Remote sensing imagery can greatly enhance the interpretation, visualisation and presentation of model outcomes, e.g., by providing a recognisable background to the spatial patterns produced by the model. Furthermore, dynamic spatial urban models provide improved abilities to assess future growth and to create planning scenarios, allowing us to explore the impacts of decisions that follow different urban planning and management policies (Klosterman 1999). Yet, the application and performance of the models is still limited by the quality and scope of the data needed for their parameterisation, calibration, and validation. Furthermore, urban modelling still suffers from a lack of knowledge and understanding of the physical and socioeconomic drivers that contribute to the pattern and dynamics of urban areas (Longley and Mesev 2000; Batty and Howes 2001; Herold et al. 2003a; Dietzel et al. 2005). In this context, Longley and Tobon (2004) emphasise that extending the interests of urban geographers towards more direct, timely, spatially disaggregate urban indicators is key in developing the data foundations to a new, data rich and relevant urban geography.

Some other references that may aid in studies related to modelling of urban growth and sprawl are furnished as follows. Worth mentioning that although the following studies are not based on remote sensing data but their applicability can be extended to remote sensing data.

Kim et al. (2006)—extracted spatial rules using decision tree method, which were then applied for urban growth modelling based on CA.

Deal and Schunk (2004)—demonstrated economic impact analysis submodel developed within the Land-use Evolution and Impact Assessment Modelling (LEAM) environment for the assessment of the cost of urban sprawl in Kane County, Illinois.

Antoni (2001)—presented a spatio-temporal methodological approach for urban sprawl modelling. The approach consists of three steps, each one corresponds to a model: the first step deals with the quantification of the sprawl (transition model), the second with the location of sprawl (potential model), and the third one with defining the land-use category of a cell (CA).

Chapter 8
Limitations of Urban Growth Analysis

8.1 Introduction

This chapter is aimed to discuss several limitations of urban growth analysis in brief. Remote sensing data are challenged by spatial, spectral, radiometric, and temporal resolutions as explained in Chap. 4. In addition to these general resolution constrains, there are several other issues associated with growth measurement and analysis.

The development and evaluation of a framework for spatial metrics analyses of datasets derived from urban remote sensing must deal with both theoretical and methodological problems. These relate to issues of scale in the selection and analysis of appropriate remote sensing data and in the application of the spatial metrics, and to the selection of the appropriate spatial metrics themselves. Herold et al. (2005b) documented these issues.

Spatial statistics also confronts many fundamental issues in the definition of its objects of study, in the construction of the analytic operations to be used, in the use of computers for analysis, in the limitations and particularities of the analyses which are known, and in the presentation of analytic results. When results obtained from spatial statistics are presented as maps, the presentation combines the spatial data which may be grossly inaccurate. Some of these issues have been discussed by Monmonier (1996).

8.2 Data and Scale Dependency

Data limitations are a primary reason for the lack of fine-scale, spatially explicit analysis of urban systems. Land-use/land-cover data that are generated from remotely sensed imageries are highly dependent on the reliability of the raw images in terms of accuracy. As a result of different data generation methods, resultant datasets can vary widely in their definitions, spatial resolution and overall accuracy, and not all datasets are appropriate for quantifying fine-scale variations in urban patterns.

In this connection it is worth describing the scale dependency. There are mainly three different scale dependencies—(1) spatial resolution of the data (spatial scale), (2) the scale at which patterns are quantified, and (3) the scale at which patterns are summarised.

8.2.1 Spatial Scale

An important consideration of remote sensing data is the spatial detail or spatial resolution in urban growth and/or pattern analysis. Data accuracy and resolution directly affect landscape heterogeneity as represented in the mapping product and determine the appropriate spatial scale of the investigation. This issue is central to all remote sensing data analysis and has been recognised in related research (Woodcock and Strahler 1987). The lower the spatial resolution, the more generalised the structure of the mapped features (e.g. urban land-cover objects) and their spatial heterogeneity will be in both the image data and the metrics. In a low spatial resolution image, individual objects may appear artificially compact or they may get merged together (Fig. 8.1). The spatial measures are then dominated more by the shape of the pixels rather than by the actual object-shape of interest (Krummel et al. 1987; Milne 1991). Furthermore, specific kinds of structures, especially linear features, may not be represented at all, thus leading to an overestimation of landscape homogeneity (Fig. 8.1). In an area of low-density development where houses are relatively far apart, a spatial resolution of 30 m will produce an estimate of developed land four times that produced using the same underlying data but a spatial resolution of 15 m.

Apparently, the most preferred spatial data are those that are sufficiently fine scale to represent individual units, e.g., individual land parcels or houses. Important

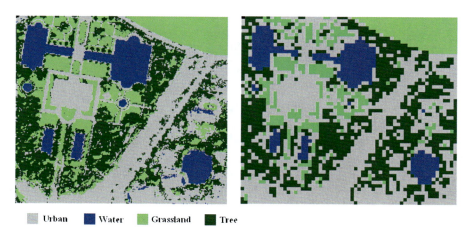

Fig. 8.1 Effects of changing pixel size on a landscape map (the *left* and *right* images are of 2.6 and 10 m spatial resolutions respectively)

to realise that although higher spatial resolution provides better interpretability by a human observer but a very high resolution leads to a high object diversity which may end up in problems when a classification algorithm is applied to the data; or it may produce a very high number of patches resulting in complications in metric analysis. Due to the increased heterogeneity in high resolution images analysis of spatial association or spatial heterogeneity will also be influenced at a high degree.

Ultimately, the question—what is the optimal spatial scale, still remains ambiguous. This ambiguity is another problem that precludes the selection of appropriate remote sensing imagery. Besides the spatial resolution, the scale of internal discrimination or subdivision of the study area is one of the central issues of spatial scale in metric analysis (Herold et al. 2005b). This issue is addressed in Sect. 8.6 in a greater detail.

8.2.2 Pattern Quantification Scale

The spatial location of land-use/land-cover classes relative to each other can be quantified using a variety of pattern measures, each of which is defined for a given spatial extent. For example, two basic approaches to measuring pattern are those that quantify the pattern surrounding individual units of observation (e.g., a pixel) and those that quantify the shape of patches defined by homogeneous and contiguous units (pixels) aggregated in space (e.g., a patch that is comprised of many contiguous and homogeneous pixels). The former rely critically on the extent of the neighbourhood surrounding individual units of observation; the latter rely critically on the extent of the area within which the patches are measured.

An important distinction between these two approaches is that the individual-level measures generate as many observations on pattern as there are observations in the dataset. In contrast, a patch-level analysis often relies on a moving window of a fixed size and therefore the number of observations is determined by the size of the analysis window relative to the size of the study region. Both are clearly dependent on the spatial resolution of the data and can give varying results. For example, using remote sensing imagery with a spatial resolution of 60 m makes it more likely that small clusters of homogeneous classes will be classified as isolated cells; in comparison, measure from 15 m resolution image likely to appear more scattered. On the other hand, aggregation (resampling) from 15 to 90 m resolution will also cause the pattern at the patch level to become simplified with fewer edges and a higher area-to-perimeter ratio, revealing a more compact urban pattern.

The more fundamental issue of scale requires ensuring that the conclusion of the analysis does not depend on any arbitrary scale. Landscape ecologists failed to do this for many years and for a long time characterised landscape elements with quantitative metrics are depended on the scale at which they were measured. They eventually developed a series of scale invariant metrics.

8.2.3 Pattern Summarisation Scale

The third scale, the scale at which patterns are summarised, is reasonable to describe the variation in pattern at the same spatial scale at which it is measured. It is sometimes useful to summarise the average pattern or variation in pattern at a coarser spatial scale. For example, Burchfield et al. (2006) measured residential pattern using 30 m cell size and summarised it at a more aggregate level by taking the average value across all residential cells within a metro region. Because pattern measures often exhibit substantial correlations and are not necessarily normally distributed, the aggregation results are often dependent on the scale at which the pattern is described. In addition, aggregating pattern to a regional level, e.g., multi-county or state, can mask important variations across or within counties that are critical to measuring and modelling residential patterns and their evolution over time.

8.3 Data Generation Methods

Data generation method includes remote sensing imaging, post-imaging processing (e.g., image classification), and generation of ancillary data. Different sensors generate different type of images that are often to be considered in temporal analysis that result in different images to be compared. Different image processing technique (or algorithm) may also generate different results for the same image. Different classification schemes (level of classification) also generate different resultant maps. Further, the generation method of ancillary data that are often required for analytical or other purposes (e.g., validation) may also vary in a wide spectrum. Use of these varying data generation methods is a common practice in urban analysis; because data generation methods and resulting data are not designed to be consistent with different datasets (Homer et al. 2004). However, the question arises regarding the validity of directly comparing spatial data derived from different data generation processes to determine the magnitude, location and pattern of development. Depending on the data generating process used (for example, the spatial resolution, spatial accuracy, classification schemes and the rules used to define various classes), data may or may not be comparable.

8.4 Classification Accuracy

Although classified satellite imagery at a range of scales has long been used in urban applications, classification of land-covers within an urban landscape is a difficult task. Urban areas are highly heterogeneous; generally there are many landscape features present in a small transact within the urban environment. This results in virtually every pixel of even high spatial resolution imagery being a mixture of a vast range of different surfaces (mixed pixel problem). In addition, identifying accurately which pixel matches which area on the ground can represent a difficult problem, leading to registration errors. Error and uncertainty in remote sensing based land-cover maps also represent a major drawback to operational application. Inaccuracies, which may be produced when converting spectral reflectance values to land-cover classes, are further compounded when inferring land-use from such land-cover maps. The classification capabilities of remote sensing data mainly depend on the spectral contrast between the classes of interest and the spectral resolution of the sensor. The lower the spectral separability of classes to be determined means less accuracy in the classified land-cover map.

Classification of remote sensing data to urban scenes is therefore fraught with difficulties, especially when attempting to segment the typically heterogeneous image structures. Thematic accuracy obviously directly influences the further analysis of the map with spatial metrics (Barnsley and Barr 2000). An overall classification accuracy of 85% is commonly considered sufficient for a remote sensing data product (Anderson et al. 1976). Post-classification comparison of two such temporal images will produce an accuracy around 72% ($0.85 \times 0.85 \times 100$), that may not be acceptable in many of the instances.

Furthermore, a generalised class definition in the classification process may result in a representation of the landscape that is too homogenous, therefore, important structural features may not be detectable with spatial metrics. In contrast, if the landscape classification is too detailed, relevant structures may get lost in a highly heterogeneous pattern (Herold et al. 2005b). Important to realise, the classification accuracy of the remote sensing data usually decreases as more classes are derived.

8.5 Selection of Metrics

A number of different approaches in representing spatial concepts have resulted in the development of various spatial metrics or metric categories as descriptive statistical measurements of spatial structures and patterns. Also for the quantification and characterisation a number of metrics have been practiced by the researchers.

Gustafson (1998) suggested for the interpretation, analysis, and evaluation of these metrics for their ability to capture the thematic information of interest for a specific application. Geoghegan et al. (1997), Alberti and Waddell (2000), Parker et al. (2001) and Herold et al. (2003b) suggest and compare a wide variety of different metrics. However, their comparison has not suggested any standard set of metrics best suited for use in urban environments as the significance of specific metrics varies with the objective of the study and the characteristics of the urban landscape under investigation. The metrics chosen for analysis must be selected based on the objectives of the study.

Important to mention, many metrics are correlated with one another, containing redundant information. Riitters et al. (1995) examined the correlations among 55 different landscape metrics by factor analysis and identified only five independent factors. Thus, many typical landscape metrics do not measure different qualities of spatial pattern. The analyst should select metrics that are relatively independent of one another, with each metric (or grouping of metrics) able to detect meaningful structure of urban landscape. It is often necessary to have more than one metric to characterise an urban landscape because one metric can not say about all. However, use of many metrics results in many measures that are often difficult to interpret resulting in difficulties for reaching to a black-and-white conclusion. Use of highly correlated metrics does not yield new information, but makes interpretations more difficult. 'Just because something can be computed does not mean that it should be computed' (Turner et al. 2001).

Another challenge is that the statistical properties and behaviour of some metrics are not well known (Turner et al. 2001). In cases where a single number is reported for a landscape, we may have little understanding of the degree to which landscape pattern must change to be able to detect significant change in the numeric value of the metric. Therefore, the analyst should definitely consider the criteria that will be applied to determine whether the observed change is meaningful or not.

Furthermore, metrics that quantify composition of landscape are not usually spatially explicit; for example the proportion of built-up area to the area of the city. This type of metrics measures what is present and in their relative amounts, or properties, without reference to where on the landscape they may be located. Therefore, the analyst should be aware about the spatial/non-spatial nature of the selected metrics.

8.6 Definition of the Spatial Domain

A basic problem in the application of both spatial metrics and spatial statistics is the definition and spatial discrimination of spatial entities for quantification and analysis. This section addresses the problems associated with spatial metrics. In general, metrics can characterise structures or features of an individual patch as a spatially and thematically consistent area representing an elementary landscape element (McGarigal et al. 2002). Metrics can also describe properties of patch classes (e.g. sum or mean values of individual patch metrics), and some can summarise properties of the entire landscape (e.g. the contagion metric) (Herold et al. 2005b). It is always important to define the *spatial domain* of the study as it directly influences the spatial metrics. Spatial domain, in general, refers to the geographic extent under analysis and its sub-divisions. In some studies the extent of the study area determines the spatial domain. For others, in particular in the comparative evaluation of intra-urban structures, it is essential to decompose the urban environment into relatively homogenous units that will serve as the spatial domains of the metric analysis.

Determining the extent of study area is not an easy task. Natural morphological urban boundary is a better consideration to designate the spatial domain in view of the dynamics of urban growth. However, the definition of natural urban extent still remains problematic (refer Sect. 1.4) and individual studies must determine their own rules for differentiating urban from rural land (Herold et al. 2003a). Another problem with morphological boundary is the lack of ancillary data that we often require in urban growth analysis and modelling. These boundaries rarely have the data on socioeconomic variables. In most of the countries, data on these variables are available in respect of administrative boundaries (or *census tracts*). As a result, researchers depend also on the administrative boundary or its different forms (Knox 1994; Batisani and Yarnal 2009; Bhatta 2009a; Martinuzzi et al. 2007; Hanham et al. 2009).

Several studies have also considered a circular area from the city centre to demarcate the spatial domain of the analysis (Kumar et al. 2007; Bhatta et al. 2010a). Buffered zones around road network (Yeh and Li 2001a) or central urban core (Herold et al. 2005a; Yeh and Li 2001) have also been considered in some of the studies. It is important to mention that the spatial discrimination and thematic definition of the spatial units must consider the characteristics of the landscape, the objectives of the study, and the use of the metrics in further analysis that may require a specific spatial subdivision of the study area.

Other than the approaches to define the spatial extent, there are many different ways of spatially subdividing an urban region based on administrative or natural/morphological or other boundaries, remote sensing and/or map analysis, or on urban modelling considerations. A common way is through the use of a regular grid as used in many urban models (Landis and Zhang 1998; Pijankowski et al. 1997). A similar concept in remote sensing data analysis is the quadratic window or kernel used to analyse features in the neighbourhood of a pixel. The neighbourhood is determined by the size of the moving kernel and its spectral or thematic characteristics are derived statistically. Barnsley and Barr (2000) reported several problems related to kernel-based approaches in urban analysis. For example: 'grid-based approaches tend to smooth the boundaries between discrete land-cover/land-use parcels; it is difficult to determine a priori the optimum kernel size; and, a rectangular window represents an artificial area that does not conform to real parcels or land-use units, which tend to have irregular shapes and their own distinct spatial boundaries' (Herold et al. 2005b). On the other hand, region-based approaches allow the discrete characterisation of thematically and functionally defined areas that are generally irregularly shaped (Barnsley and Barr 2000; Barr and Barnsley 1997; Gong et al. 1992).

Needless to mention, regional subdivisions of urban space may vary extensively in size, shape, and purpose. Governmental and planning organisations generally use census tracts and blocks or zoning districts, based on the characteristics of the built environment, socioeconomic variables, administrative boundaries and other considerations (Knox 1994). Urban models may use a wide variety of spatial units, including individual parcels associated with key human agents such as landowners participating in micro-economic processes (Irwin and Geoghegan 2001; Waddell 1998), and uniform analysis zones defined by the multiple intersections of polygons on different data layers representing natural and socioeconomic variables of interest (Klosterman 1999). The definition of regions based on remote sensing data uses automated, semiautomated or supervised approaches. Automated techniques are usually based on pattern recognition or image segmentation, which result in areas with similar spectral and textural pattern. A traditional approach in region-based remote sensing analysis is the concept of photomorphic region developed for aerial photographic interpretation (Peplies 1973). Photomorphic regions are defined as image segments with similar properties of size, shape, tone/colour, texture and pattern. Circular study areas are often subdivided in concentric zones and/or pie sections (Kumar et al. 2007; Bhatta et al. 2010a). Barr and Barnsley (1997) following Barnsley et al. (1995) discuss a combined remote sensing and GIS approach for deriving urban morphological zones that describes the physical extent of the built up area based on remote sensing data, modified by criteria of minimum size and spatial contiguity based on GIS data. In general, all these approaches are appropriate for spatial metrics analysis in urban environments, but region-based sub-division methods are likely to provide a better segmentation of urban space for most applications (Herold et al. 2005b).

The scale of internal discrimination or subdivision of the study area is another key issue in metric analysis that actually relates to spatial scale (Sect. 8.2.1). The available approaches can overcome the averaging nature of metrics over an entire study area that may lead to incorrect interpretations of the dynamics in the region. For example, as stated by Herold et al. (2005b), changes reflected in the spatial metrics cannot usually be related to specific locations within the urban area without visual spatial interpretation or some more detailed analysis at the patch level. Furthermore, temporal variations in the spatial metrics may result from the aggregate or cumulative effects of different dynamic processes. Spatial desegregation allows the study area to be considered as a set of smaller individual landscapes and regionalises the metric analysis to an appropriate scale. An approach may be to consider the sub-divisions at different scales; e.g., the entire city, borough boundary, and ward boundary. Even so, it may be impossible to directly relate the metric changes to specific urban change processes. For studies of structural change in urban land-use, a definition of more or less homogenous urban land-use units are usually developed before the analysis can begin. These have to be defined and spatially differentiated using the available data sources (e.g. remote sensing or/and census data) and any other relevant information and local knowledge (Bhatta 2009a).

Similar problems, as discussed in this section, are also associated with spatial statistics as well; where they are commonly referred to as *modifiable areal unit problem* (refer Sect. 8.10).

8.7 Spatial Characterisation

Spatial statistical techniques favour the spatial definition of objects as points because there are very few statistical techniques which can operate directly on line, area, or volume elements. One can imagine the problem of defining every geographic object as points. Computer tools, on the other hand, favour the spatial definition of objects as homogeneous and separate elements because of the primitive nature of the computational structures available and the ease with which these primitive structures can be created. There may also be arbitrary effects introduced by the spatial bounds or limits placed on the phenomenon or study area. This occurs since spatial phenomena may be unbounded or have ambiguous transition zones. This creates edge effects from ignoring spatial dependency or interaction outside the study area. It also imposes artificial shapes on the study area that can affect apparent spatial patterns such as the degree of clustering. A possible solution is similar to the sensitivity analysis strategy for the modifiable areal unit problem (i.e., change the limits of the study area and compare the results of the analysis under each realisation). Another possible solution is to overbound the study area. It is also feasible to eliminate edge effects in spatial modelling and simulation by mapping the region to a boundless object such as a sphere.

8.8 Spatial Dependency (Autocorrelation)

A fundamental concept in geography is that nearby entities often share more similarities than entities which are far apart (Miller 2004). This idea is often labelled 'Tobler's *first law of geography*' and may be summarised as 'everything is related to everything else, but near things are more related than distant things' (Tobler 1970).

Spatial dependency is the co-variation of properties within a geographical space; i.e., characteristics at proximal locations appear to be correlated, either positively or negatively. There are at least three possible explanations of this. One possibility is there is a simple spatial correlation relationship: whatever is causing an observation in one location also causes similar observations in nearby locations. For example, the causes that support migration of a farmer to a city are also responsible to attract other farmers in the metropolitan area. Another possibility is spatial causality: something at a given location directly influences the characteristics of nearby locations. For example, conditions that support construction within an area will also have influence on neighbouring areas and these neighbouring areas will also experience new construction. A third possibility is spatial interaction: the movement of people, goods or information creates apparent relationships between locations. For example, suburban growth occurs due to accessibility of urban core through road networks to the rural area.

Spatial dependency leads to the spatial autocorrelation problem in statistics; since this violates standard statistical techniques that assume independence among observations. For example, regression analyses that do not compensate for spatial dependency can have unstable parameter estimates and yield unreliable significance tests. However, spatial regression models generally capture these relationships and do not suffer from these weaknesses.

Spatial effects also manifest as spatial heterogeneity, or the apparent variation in a process with respect to location in a geographical space. Unless a space is uniform and boundless, every location will have some degree of uniqueness relative to the other locations. This affects the theory of spatial dependency relations and therefore the spatial processes.

8.9 Spatial and Temporal Sampling

Spatial sampling means determining a limited number of locations in a geographical space. These samples are then used to estimate the unknown values at neighbouring locations. This approach, however, is subject to be affected by spatial dependency and heterogeneity. Spatial dependency suggests that since one location can predict the value of another location, we do not need observations in both places. But heterogeneity suggests that this relation can change across space, and therefore we cannot trust an observed degree of dependency beyond a region that may be very small. Basic spatial sampling schemes include random, clustered and systematic sampling. These basic schemes can be applied at multiple levels in a designated spatial hierarchy (e.g., urban area, city, and neighbourhood). It is also possible to exploit ancillary data, for example, using property values as a guide in a spatial sampling scheme to measure educational attainment and income.

Similar to spatial sampling, *temporal sampling* (*temporal frequency*) assumptions are also affected by temporal dependency and temporal heterogeneity. Often we assume that geographical relations flow through time. For example, as a temporal dependency, urban growth rate or population growth rate of an area likely to be continued in near future. This assumption helps to simulate several phenomena including urban growth and sprawl. However, there may be several natural or other factors that may interrupt this process and may result in temporal heterogeneity. This possibility increases if the temporal samples that have been considered in simulation and modelling are far apart.

8.10 Modifiable Areal Unit Problem

The *modifiable areal unit problem* (MAUP) is an issue of spatial statistics in the analysis of spatial data arranged in zones, where the conclusion depends on the particular shape or size of the zones used in the analysis (Openshaw 1984; Openshaw and Alvanidies 1999). Spatial analysis and modelling often involves aggregation of spatial units such as census tracts or traffic analysis zones. These units may reflect data collection and/or modelling convenience rather than homogeneous, cohesive regions in the real world. The spatial units are therefore arbitrary or modifiable and contain artefacts related to the degree of spatial aggregation or the placement of boundaries.

The effects of the MAUP can be divided into two components: the *scale effect* and the *zoning effect* (Armhein 1995). The *scale effect* is the variation in numerical results that occurs due to the number of zones used in an analysis. The *zoning effect* is the variation in numerical results arising from the grouping of small areas into larger units. Figure 8.2a shows the change in means that occurs when smaller units are aggregated into larger units, and Fig. 8.2b shows the change in means with the change of zoning system.

The MAUP arises in urban analysis due to the fact that an infinite number of zoning systems could be constructed to subdivide a city into smaller areal units; this

8.10 Modifiable Areal Unit Problem

a) Scale effect

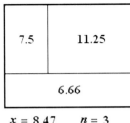
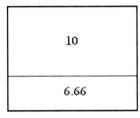

b) Zoning effect

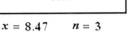
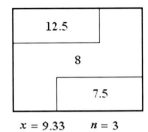

x: mean n: number of zones

Fig. 8.2 The MAUP effects

implies that the data reported for areal units will differ between zoning systems. Therefore, the results will vary for the same analytical technique, but different zoning systems for the same geographical area (please refer Wong et al. 1999; Horner and Murray 2002). Worth mentioning that the aggregation of point data into zones of different shapes and sizes may even lead to opposite conclusions.

The seriousness of MAUP issue is clearly demonstrated by the wide variety of techniques whose results are known to be affected by it. These techniques include *correlation analysis* (Gehlke and Biehl 1934; Blalock 1964; Openshaw and Taylor 1979), *regression analysis* (Fotheringham and Wong 1991; Amrhein and Flowerdew 1992; Amrhein 1995; Okabe and Tagashira 1996; Tagashira and Okabe 2002), *spatial interaction modelling* (Openshaw 1977b; Batty and Sikdar 1982a,b,c,d; Putman and Chung 1989), *location-allocation modelling* (Goodchild 1979; Fotheringham et al. 1995; Hodgson et al. 1997; Murray and Gottsegen 1997) and *discrete choice modelling* (Guo and Bhat 2004). All of which are commonly used in urban analysis. Furthermore, the MAUP has been shown to affect indices derived from areal data such as the segregation index (Wong et al. 1999) and the excess commute (Horner and Murray 2002). Indices such as these are also used in urban analysis.

The MAUP has two components: the resolution effect and the zoning effect (Páez and Scott 2004). The resolution effect is a consequence of spatial aggregation within a pixel—that is, results from the same analytical technique tend to vary according to the level of spatial resolution. Horner and Murray (2002) have shown that estimates of excess commuting decrease with increasing pixel size. Other studies have documented increases in the values of correlation coefficients as spatial resolution becomes coarser (Gehlke and Biehl 1934; Blalock 1964; Openshaw and Taylor 1979). The zoning effect, on the other hand, results from the multitude of zoning schemes that could be constructed and used at any given scale. Several studies have documented this effect (e.g., Fotheringham and Wong 1991; Fotheringham et al. 1995; Wong et al. 1999; Horner and Murray 2002), which basically show that for a given level of spatial resolution there exists a range of possible outcomes from an analytical technique owing to changes in zoning.

The *frame-independence*[1] can be achieved in urban analysis by the use of individual-level data (Fotheringham 1989; Tobler 1989). Such data differ from areal data; and their locations are represented by geographical coordinates, rather than zones. This implies that the objects or phenomena under scrutiny are visualised as points in space. Trips, individuals, households, firms and land parcels are but a few objects in a city for which individual-level data are routinely collected, especially in developing countries. As a result, such data are difficult to obtain to support basic urban geographic research (Fotheringham et al. 2000). Confidentiality is another major issue for individual-level data, which must be preserved at all costs (Páez and Scott 2004). This implies that if the data are released in public domain, locational precision is often sacrificed. At best, locations are represented by zones, for example, census data are not supplied for each household rather they are represented by zones.

There are several methods available to mitigate MAUP effects in urban analysis—e.g., consideration of smallest areal units (Wong 1996; Fotheringham et al. 2000), analysis over a wide range of zoning systems (Fotheringham et al. 2000; Fotheringham and Wong 1991; Murray and Gottsegen 1997; Horner and Murray 2002; Openshaw and Baxter 1977; Openshaw 1977a, b, 1978a, b; Openshaw and Rao 1995), adjusting the aggregate-level variance-covariance matrix (Steel and Holt 1996; Holt et al. 1996; Tranmer and Steel 1998), amongst others. However, finally, what works at best still remains ambiguous. Researchers have made their own assumptions and considerations to encounter MAUP effects.

---

[1] Frame-independence means that the results from the analysis do not depend on the zoning system used.

References

Abed, J. and Kaysi, I. (2003). Identifying urban boundaries: application of remote sensing and geographic information system technologies. *Canadian Journal of Civil Engineering*, **30**, 992–999.

Abrahart, R.J. and Kneale, P.E. (1997). Exploring neural network rainfall-runoff modeling. *Proceedings of 6th British Hydrological Society Symposium*, pp. 935–944.

Acevedo, W., Foresman, T.W. and Buchanan, J.T. (1996). Origins and philosophy of building a temporal database to examine human transformation processes. *Proceedings of ASPRS/ACSM Annual Convention and Exhibition*, Baltimore, MD, April 22–24, vol. 1, pp. 148–161. URL: http://landcover.usgs.gov/urban/umap/pubs/asprs_wma.php.

Acioly, C.C. and Davidson, F. (1996). Density in urban development. *Building Issues*, **8**(3), 3–25.

Agarwal, C., Green, G.L., Grove, M., Evans, T. and Schweik, C. (2000). *A Review and Assessment of Landuse Change Models: Dynamics of Space, Time, and Human Choice*. US Forest Service and Center for the Study of Institutions, Population, and Environmental Change (CIPEC), Indiana University, USA. URL: http://hero.geog.psu.edu/archives/AgarwalEtALInPress.pdf.

Aguayo, M.I., Wiegand, T., Azócar, G.D., Wiegand, K., and Vega, C.E. (2007). Revealing the driving forces of mid-cities urban growth patterns using spatial modeling: a case study of Los Ángeles, Chile. *Ecology and Society*, **12**(1). URL: http://www.ecologyandsociety.org/vol12/iss1/art13/.

Al-Ahmadi, K., Heppenstall, A., Hogg, J. and See, L. (2008a). A fuzzy cellular automata urban growth model (FCAUGM) for the city of Riyadh, Saudi Arabia, part 2: scenario testing. *Applied Spatial Analysis and Policy Journal*, doi:10.1007/s12061-008-9019-z.

Al-Ahmadi, K., Heppenstall, A., Hogg, J. and See, L. (2009). A fuzzy cellular automata urban growth model (FCAUGM) for the city of Riyadh, Saudi Arabia. part 1: model structure and validation. *Applied Spatial Analysis*, **2**, 65–83.

Al-Ahmadi, K., See, L., Heppenstall, A. and Hogg, J. (2008b). Calibration of a fuzzy cellular automata model of urban dynamics in Saudi Arabia. *Ecological Complexity*, doi:10.1016/j.ecocom.2008.09.004.

Alberti, M. and Waddell, P. (2000). An integrated urban development and ecological simulation model. *Integrated Assessment*, **1**, 215–227.

Alberti, M., Weeks, R. and Coe, S. (2004). Urban land-cover change analysis in central Puget Sound, *Photogrammetric Engineering and Remote Sensing*, **70**, 1043–1052.

Alexander, C. (1965). A city is not a tree. *Architectural Forum*, **122**(1), 58–62.

Alkheder, S. and Shan, J. (2005). Cellular automata urban growth simulation and evaluation—a case study of Indianapolis. *Proceedings of 8th International Conference on GeoComputation*, University of Michigan., USA, 31 July–3 August. URL: http://www.geocomputation.org/2005/index.html.

Allen, J. and Lu, K. (2003). Modeling and prediction of future urban growth in the Charleston region of South Carolina: a GIS-based integrated approach. *Conservation Ecology*, **8**(2). URL: http://www.consecol.org/vol8/iss2/art2/.

References

Allen, P.M. (1997). *Cities and Regions as Self-organizing Systems: Models of Complexity*. Taylor and Francis, London.

Almeida, C.M., Monteiro, A.M.V., Mara, G., Soares-Filho, B.S., Cerqueira, G.C., Pennachin, C.S.L. and Batty, M., (2005). GIS and remote sensing as tools for the simulation of urban land-use change. *International Journal of Remote Sensing*, 26(4), 759–774.

Alperovich, G. and Deutsch, J. (2002). An application of a switching regimes regression to the study of urban structure. *Papers in Regional Science*, 81, 83–98.

Alphan, H. (2003). Land use change and urbanization in Adana, Turkey. *Land Degradation and Development*, 14(6), 575–586.

Alterman, R. (1997). The challenge of farmland preservation: lessons from a six-nation comparison. *Journal of the American Planning Association*, 63(2), 220–241.

Amrhein, C.G. (1995). Searching for the elusive aggregation effect: evidence from statistical simulations. *Environment and Planning A*, 27, 105–119.

Amrhein, C.G. and Flowerdew, R. (1992). The effect of data aggregation on a Poisson regression-model of Canadian migration. *Environment and Planning A*, 24, 1381–1391.

Anderson, H.H. (1999). *Use and Implementation of Urban Growth Boundaries*. Center for Regional and Neighborhood Action, Denver.

Anderson, J.R., Hardy, E.E., Roach, J.T. and Witmer, R.E. (1976). A land use and land cover classification system for use with remote sensor data. *US Geological Survey Professional Paper 964*. United States Government Printing Office, Washington, DC. URL: http://landcover.usgs.gov/pdf/anderson.pdf.

Anderson, W.P., Kanaroglou, P.S. and Miller, E.J. (1996). Urban form, energy and the environment: a review of issues, evidence and policy. *Urban Studies*, 33(1), 7–35.

Angel, S., Parent, J. and Civco, D. (2007). Urban sprawl metrics: an analysis of global urban expansion using GIS. *Proceedings of ASPRS 2007 Annual Conference*, Tampa, Florida May 7–11. URL: http://clear.uconn.edu/publications/research/tech_papers/Angel_et_al_ASPRS2007.pdf.

Angel, S., Sheppard, S.C., and Civco, D.L. (2005). *The Dynamics of Global Urban Expansion*. Transport and Urban Development Department, The World Bank, Washington, DC, p. 200. URL: http://www.citiesalliance.org/doc/resources/upgrading/urban-expansion/worldbankreportsept2005.pdf.

Anselin, L. (1995). Local indicators of spatial association – LISA. *Geographical Analysis*, 27, 93–115.

Angelici, L., Bryntn, A. and Friwmasn, Z. (1977). Techniques for land use change detection using Landsat imagery. *Proceedings of the American Society of Photogrammetry*, 21, 7–228.

Anselin, L. and Griffith, D.A. (1988). Do spatial effects really matter in regression-analysis. *Papers of the Regional Science Association*, 65, 11–34.

Anthony, J. (2004). Do state growth management regulations reduce sprawl? *Urban Affairs Review*, 39, 367–397.

Antoni, J. (2001). Urban sprawl modelling: a methodological approach. *Cybergeo: European Journal of Geography*. URL: http://www.cybergeo.eu/index4188.html.

Archer, R.W. (1973). Land speculation and scattered development: failures in the urban-fringe market. *Urban Studies*, 10, 367–372.

Armhein, C. (1995). Searching for the elusive aggregation effect: evidence from statistical simulations. *Environment and Planning A*, 27(1), 105–119.

Armstrong, M.P. (1988). Temporality in spatial databases. *Proceedings of GIS/LIS'88, vol. 2*, pp. 880–889.

Arnott, R. and McMillen, D.P. (2006). *A Companion to Urban Economics*. Blackwell Publishing, Malden, MA, p. 574.

Asif, S. and Shachar, A. (1999). *LAND—National Comprehensive Outline Scheme and Principles for Building and Development Document—NOS/35, Interim Report—Summery of Stage A*. Tel-Aviv, Hebrew.

Aspinall, R.J. and Hill, M.J. (1997). Land cover change: a method for assessing the reliability of land cover changes measured from remotely sensed data. *Proceedings of 1997 International Geoscience and Remote Sensing Symposium*, Singapore, August 4–8, pp. 269–271.

References

Audirac, I., Shermyen, A.H. and Smith, M.T. (1990). Ideal urban form and visions of the good life: Florida's growth management dilemma. *Journal of the American Planning Association*, **56**(4), 470–482.

Axtell, R.L. (2000). *Why Agents? On the Varied Motivations for Agent Computing in the Social Sciences*. Center on social and economic dynamics (The Brookings Institute): Working Paper 17, Washington, DC.

Axtell, R.L. and Epstein, J.M. (1994). Agent-based modeling: understanding our creations. *The Bulletin of the Santa Fe Institute*, **9**, 28–32, Winter.

Bae, C.H.C. and Richardson, H.W. (1994). *Automobiles, the Environment and Metropolitan Spatial Structure*. Lincoln Institute of Land Policy, Cambridge, MA.

Bahr, H. (2004). Image segmentation for change detection in urban environments. In: J.P. Donnay, M.J. Barnsley and P.A. Longley (eds.), *Remote Sensing and Urban Analysis*, Taylor & Francis, London, pp. 95–114.

Banister, D., Watson, S. and Wood, C. (1997). Sustainable cities: transport, energy, and urban form. *Environment and Planning B*, **24**, 125–143.

Banner, A.V. and Lynham, T. (1981). Multitemporal analysis of Landsat data for forest cut over mapping—a trial of two procedures. *Proceedings of the 7th Canadian Symposium on Remote Sensing*, Winnipeg, Manitoba, Canada, pp. 233–240.

Barnes, K.B., Morgan, J.M., III, Roberge M.C. and Lowe, S. (2001). *Sprawl Development: Its Patterns, Consequences, and Measurement*. A white paper, Towson University. URL: http://chesapeake.towson.edu/landscape/urbansprawl/download/Sprawl_white_paper.pdf.

Barnsley, M.J. and Barr, S.J. (1996). Inferring urban land use from satellite sensor images using kernel-based spatial reclassification. *Photogrammetric Engineering and Remote Sensing*, **62**, 949–958.

Barnsley, M.J. and Barr, S.L. (1997). A graph based structural pattern recognition system to infer urban land-use from fine spatial resolution land-cover data. *Computers, Environment and Urban Systems*, **21**(3&4), 209–225.

Barnsley, M.J. and Barr, S.L. (2000). Monitoring urban land use by earth observation. *Surveys in Geophysics*, **21**, 269–289.

Barnsley, M.J., Barr, S.L. and Sadler, G.J. (1995). Mapping the urban morphological zone using satellite remote sensing and GIS: UK involvement in the Eurostat pilot project on remote sensing and urban statistics, remote sensing in action. *Proceedings of Annual Conference of the Remote Sensing Society*, University of Southampton, September 11–14, Remote Sensing Society of Nottingham, pp. 209–216.

Barr, S. and Barnsley, M. (1997). A region-based, graph-oriented data model for the inference of second order information from remotely-sensed images. *International Journal of Geographical Information Science*, **11**(6), 555–576.

Barredo, J.I., Demicheli, L., Lavalle, C., Kasanko, M. and McCormick, N. (2004). Modelling future urban scenarios in developing countries: an application case study in Lagos, Nigeria. *Environment and Planning B*, **31**, 65–84.

Batisani, N. and Yarnal, B. (2009). Urban expansion in Centre County, Pennsylvania: spatial dynamics and landscape transformations. *Applied Geography*, **29**, 235–249.

Batty, M. (1976). Entropy in spatial aggregation. *Geographical Analysis*, **8**, 1–21.

Batty, M. (1998). Urban evolution on the desktop: simulation with the use of extended CA. *Environment and Planning A*, **30**(11), 1943–1967.

Batty, M. (2000). GeoComputation using cellular automata. In: S. Openshaw and R.J. Abrahart (eds.), *Geocomputation*, Taylor & Francis, New York, pp. 95–126.

Batty, M. (2001). Agent-based pedestrian modeling. *Environment and Planning B*, **28**(3), 321–326.

Batty, M. (2003). *Agents, Cells and Cities: New Representational Models for Simulating Multi-scale Urban Dynamics*. Working Paper Series, Paper 65. Center for Advanced Spatial Analysis, University College London.

Batty, M. (2005). *Cities and Complexity: Understanding Cities with Cellular Automata, Agent-Based Models, and Fractals*. The MIT Press, Cambridge, MA.

References

Batty, M. (2008). Fifty years of urban modeling: macro-statics to micro-dynamics. In: S. Albeverio, D. Andrey, P. Giordano and A. Vancheri (eds.), *The Dynamics of Complex Urban Systems: An Interdisciplinary Approach*. Springer, Berlin, pp. 1–19.

Batty, M. (2009). Urban modelling. In: R. Kitchin and N. Thrift (eds.), *International Encyclopedia of Human Geography*, Elsevier Science, London, pp. 51–58.

Batty, M. and Howes D. (2001). Predicting temporal patterns in urban development from remote imagery. In: J.P. Donnay, M.J. Barnsley and P.A. Longley (eds.), *Remote Sensing and Urban Analysis*, Taylor and Francis, London and New York, pp. 185–204.

Batty, M. and Kim, K.S. (1992). Form follows function: reformulating urban population density functions. *Urban Studies*, **29**, 1043–1070.

Batty, M. and Sikdar, P.K. (1982a). Spatial aggregation in gravity models: (1) an information-theoretic framework. *Environment and Planning A*, **14**, 377–405.

Batty, M. and Sikdar, P.K. (1982b). Spatial aggregation in gravity models: (2) one-dimensional population-density models. *Environment and Planning A*, **14**, 525–553.

Batty, M. and Sikdar, P.K. (1982c). Spatial aggregation in gravity models: (3) two-dimensional trip distribution and location models. *Environment and Planning A*, **14**, 629–658.

Batty, M. and Sikdar, P.K. (1982d). Spatial aggregation in gravity models: (4) generalizations and large-scale applications. *Environment and Planning A*, **14**, 795–822.

Batty, M. and Xie, Y. (1994). From cells to cities. *Environment and Planning B*, **21**, 31–48.

Batty, M. and Xie, Y. (1997). Possible urban automata. *Environment and Planning B*, **24**, 175–192.

Batty, M., and Longley, P. (1994). *Fractal Cities: A Geometry of Form and Function*. Academic Press, London, San Diego.

Batty, M., Desyllas, J. and Duxbury, E. (2003). Safety in numbers? Modeling crowds and designing control for the Notting Hill Carnival. *Urban Studies*. **40**(8), 1573–1590.

Batty, M., Xie, Y. and Sun, Z (1999). Modeling urban dynamics through GIS-based cellular automata. *Computers, Environment and Urban Systems*, **23**, 205–233.

Baumont, C., Ertur, C. and Le Gallo, J. (2004). Spatial analysis of employment and population density: the case of the aglomeration of Dijon 1999. *Geographical Analysis*, **36**, 146–176.

Belkina, T.D. (2007). Diagnosing urban development by an indicator system. *Studies on Russian Economic Development*, **18**(2), 162–170.

Ben Akiva, M. and Lerman, S. (1985). Discrete choice analysis: theory and application to travel demand. The MIT Press, Cambridge, MA.

Benati, S. (1997). A cellular automata for the simulation of competitive location. *Environment and Planning B*, **24**, 205–218.

Bender, B. and Hwang, H. (1985). Hedonic housing price indices and secondary employment centers. *Journal of Urban Economics*, **17**, 90–107.

Benediktsson, J.A., Swain, P.H. and Ersoy, O.K. (1990). Neural network approaches versus statistical methods in classification of multisource remote sensing data. *IEEE Transactions on Geoscience and Remote Sensing*, **28**, 540–551.

Benenson, I. (1998). Multi-agent simulations of residential dynamics in the city. *Environment and Urban Systems*, **22**(1), 25–42.

Benenson, I. and Torrens, P.M. (2004). *Geosimulation: Automata-Based Modeling of Urban Phenomena*. Wiley, New York.

Benfield, F.K., Raimi, M. and Chen, D. (1999). *Once There Were Greenfields: How Urban Sprawl Is Undermining America's Environment, Economy and Social Fabric*. The Natural Resources Defense Council, Washington, DC.

Bengston, D.N., Fletcher, J.O. and Nelson, K.C. (2004). Public policies for managing urban growth and protecting open space: policy instruments and lessons learned in the United States. *Landscape and Urban Planning*, **69**, 271–286.

Bengston, D.N., Potts, R.S., Fan, D.P. and Goetz, E.G. (2005). An analysis of the public discourse about urban sprawl in the United States: monitoring concern about a major threat to forests. *Forest Policy and Economics*, **7**, 745–756.

Bennion M.W. and O'Neill W.A. (1994). Building Transportation Analysis Zones Using Geographic Information Systems. *Transportation Research Record*, **1429**, 49–56.

References

Bento, A.M., Franco, S.F. and Kaffine, D. (2006). The efficiency and distributional impacts of alternative anti-sprawl policies. *Journal of Urban Economics*, **59**, 121–141.

Bernstein, J.D. (1994). *Land Use Considerations in Urban Environmental Management*. Urban Development Division, The World Bank, Washington, DC.

Berry, D. and Plaut, T. (1978). Retaining agricultural activities under urban pressures: a review of land use conflicts and policies. *Policy Sciences*, **9**, 153–178.

Berry, M.W., Flamm, R.O., Hazen, B.C. and MacIntyre, R.L. (1996). The land-use change and analysis system (LUCAS) for evaluating landscape management decisions. *IEEE Computational Science & Engineering*, **3**(1), 24–35.

Bertaud, A. and Malpezzi, S. (1999). *The Spatial Distribution of Population in 35 World Cities: The Role of Markets, Planning and Topography*. Center of Urban Land Economics Research, The University of Wisconsin, Madison.

Bhat, C. and Zhao, H.M. (2002). The spatial analysis of activity stop generation. *Transportation Research B*, **36**, 557–575.

Bhatta, B. (2007). Quantification of confusion in LISS–III & LISS–IV data for urban land-cover classification. *Proceedings of National Conference on High Resolution Remote Sensing & Thematic Applications*, Kolkata, December 18–20, Indian Society of Remote Sensing, p. 58.

Bhatta, B. (2008). *Remote Sensing and GIS*. Oxford University Press, New York and New Delhi.

Bhatta, B. (2009a). Analysis of urban growth pattern using remote sensing and GIS: a case study of Kolkata, India. *International Journal of Remote Sensing*, **30**(18), 4733–4746.

Bhatta, B. (2009b). Modelling of urban growth boundary using geoinformatics. *International Journal of Digital Earth*, **2**(4), 359–381.

Bhatta, B., Saraswati, S. and Bandyopadhyay, D. (2010a). Quantifying the degree-of-freedom, degree-of-sprawl, and degree-of-goodness of urban growth from remote sensing data. *Applied Geography*, **30**(1), 96–111.

Bhatta, B., Saraswati, S. and Bandyopadhyay, D. (2010b). Urban Sprawl Measurement from Remote Sensing Data. *Applied Geography*, doi 10.1016/j.apgeog.2010.02.002.

Bhuyan, M.R., Rajak, D.R. and Oza, M. (2007). Quantification of improvements by using AWiFS over WiFS data. *Journal of the Indian Society of Remote Sensing*, **35**(1), 43–52.

Bischof, H., Shneider, W. and Pinz, A.J. (1992). Multispectral classification of Landsat-images using neural networks. *IEEE Transactions on Geoscience and Remote Sensing*, **30**, 482–490.

Bishop, I. (1994). *Comparing Regression and Neural Net Based Approaches to Modeling of Scenic Beauty*. Centre for Geographic Information Systems and Modeling, The University of Melbourne, Australia.

Black J.A. (1981). *Urban Transport Planning*. Croom Helm, London.

Black, T.J. (1996). The economics of sprawl. *Urban Land*, **55**(3), 6–52.

Blalock, H. (1964). *Causal Inferences on Nonexperimental Research*. University of North Carolina Press, Chapel Hill.

Bolduc, D., Laferriere, R. and Santarossa, G. (1995). Spatial autoregressive error components in travel flow models: an application to aggregate mode choice. In: L. Anselin and R.J.G.M. Florax (eds.), *New Directions in Spatial Econometrics*, Springer-Verlag, Berlin, pp. 96–108.

Boyce, R.R. (1963). Myth versus reality in urban planning. *Land Economics*, **39**(3), 241–251.

Breheny, M. (1992). *Sustainable Development and Urban Form*. Pion, London.

Breslaw, J.A. (1990). Density and urban sprawl: comment. *Land Economics*, **66**(4), 464–468.

Brivio, P.A. and Zilioli, E. (2001). Urban pattern characterization through geostatistical analysis of satellite images. In: J.P. Donnay, M.J. Barnsley and P.A. Longley (eds.), *Remote Sensing and Urban Analysis*, Taylor and Francis; London, pp. 40–53.

Brookes, C.J. (1997). A parameterized region-growing programme for site allocation on raster suitability maps. *International Journal of Geographical Information Science*, **11**(4), 375–396.

Brown, D.G., Riolo, R., Robinson, D.T., North, M. and Rand, W. (2005). Spatial process and data models: toward integration of agent-based models and GIS. *Journal of Geographical Systems*, **7**(1), 25–47.

Brown, M., Lewis, H.G. and Gunn, S.R. (2000). Linear spectral mixture models and support vector machines for remote sensing. *IEEE Transactions on Geoscience and Remote Sensing*, **38**, 2346–2360.
Brown, T.R. and Bonifay, C. (2001). Is new urbanism the cure? A look at central Florida's response. *Real Estate Issues*, **26**(3), 21–27.
Brueckner, J.K. (1997). Infrastructure financing and urban development: the economics of impact fees. *Journal of Public Economics*, **66**, 383–407.
Brueckner, J.K. (2000). Urban sprawl: diagnosis and remedies. *International Regional Science Review*, **23**(2), 160–171.
Brueckner, J.K. (2001). Urban sprawl: lessons from urban economics. In: W.G. Gale and J.R. Pack (eds.), *Brookings-Wharton Papers on Urban Affairs*, Brookings Institution, Washington, DC, pp. 65–89.
Brueckner, J.K. and Kim, H. (2003). Urban sprawl and the property tax. *International Tax and Public Finance*, **10**, 5–23.
Brueckner, J.K. and Lai, F.C. (1996). Urban growth with resident landowners. *Regional Science and Urban Economics*, **26**, 125–143.
Brunsdon, C., Fotheringham, A.S. and Charlton, M. (1999). Some notes on parametric significance tests for geographically weighted regression. *Journal of Regional Science*, **39**, 497–524.
Brunsdon, C., Fotheringham, A.S. and Charlton, M.E. (1996). Geographically weighted regression: a method for exploring spatial nonstationarity. *Geographical Analysis*, **28**, 281–298.
Bruzzone, L. and Prieto, D.F. (2000). Automatic analysis of the difference image for unsupervised change detection. *IEEE Transactions on Geoscience and Remote Sensing*, **38**, 1171–1182.
Bruzzone, L., Prieto, D.F. and Serpico, S.B. (1999). A neural-statistical approach to multitemporal and multisource remote-sensing image classification. *IEEE Transactions on Geoscience and Remote Sensing*, **37**, 1350–1359.
Buchan, G.M. and Hubbard, N.K. (1986). Remote sensing in land-use planning: an application in west central Scotland using SPOT-simulation data. *International Journal of Remote Sensing*, **7**, 767–777.
Buiton, P.J. (1994). A vision for equitable land use allocation. *Land Use Policy*, **12**(1), 63–68.
Buliung, R.N. and Kanaroglou, P.S. (2006). Urban form and household activity travel behavior. *Growth and Change*, **37**, 172–178.
Burby, R.J., Nelson, A.C., Parker, D. and Handmer, J. (2001). Urban containment policy and exposure to natural hazards: is there a connection? *Journal of Environmental Planning and Management*, **44**(4), 475–490.
Burchell, R.W., Downs, A., McCann, B. and Mukherji, S. (2005). *Sprawl Costs: Economic Impacts of Unchecked Development*. Island Press, Washington, DC.
Burchell, R.W., Shad, N.A., Lisotkin, D., Phillips, H., Downs, A., Seskin, S., et al. (1998). *The Costs of Sprawl Revisited*. National Academy Press, Washington, DC.
Burchell, R.W., Listokin, D. and Galley, C.C. (2000a). Smart growth: More than a ghost of urban policy past, less than a bold new horizon. *Housing Policy Debate*, **11**, 821–879.
Burchell, R.W., Lowenstein, G., Dolphin, W.R., Galley, C.C., Downs, A., Seskin, S. and Moore, T. (2000b). The benefits of sprawl. In: *The Costs of Sprawl-Revisited*. Transportation Research Board and National Research Council, Washington, DC, pp. 351–391.
Burchfield, M., Overman, H., Puga, D. and Turner, M. (2006). Causes of sprawl: a portrait from space. *Quarterly Journal of Economics*. 121(2), 587–633, May 2006.
Burgess, E.W. (1925). The growth of the city: an introduction to a research project. In: R.E. Park, E.W. Burgess, R. McKenzie (eds.), *The City*, University of Chicago Press, Chicago, pp. 47–62.
Byrne, G.F., Crapper, P.F. and Mayo, K.K. (1980). Monitoring land-cover by principal components analysis of multitemporal Landsat data. *Remote Sensing of Environment*, **10**, 175–184.
Caglioni, M., Pelizzoni, M. and Rabino, G.A. (2006). Urban sprawl: a case study for project gigalopolis using SLEUTH model. In: S. El Yacoubi, B. Chopard and S. Bandini (eds.), *ACRI 2006, LNCS 4173*, Springer, pp. 436–445.
Cakir, H.I., Khorram, S. and Nelson, S.A.C. (2006). Correspondence analysis for detecting land cover change. *Remote Sensing of Environment*, **102**, 306–317.

References

Calavita, N. and Caves, R. (1994). Planners' attitudes toward growth: a comparative case study. *Journal of the American Planning Association*, 60, 483–500.

Calthorpe, P., Fulton, W. and Fishman, R. (2001). *The Regional City: Planning for the End of Sprawl*. Island Press, Washington, DC.

Candau, J.C. and Goldstein, N. (2002). Multiple scenario urban forecasting for the California south coast region. *Proceedings of 40th Annual Conference of the Urban and Regional Information Systems Association*, Chicago, IL, October 26–30.

Carter, H. (1981). *The Study of Urban Geography*. Edward Arnold Victoria, Australia.

Castle, C.J.E. and Crooks, A.T. (2006). *Principles and concepts of agent-based modeling for developing geospatial simulations*. Working Paper 110, Centre for Advanced Spatial Analysis, University College London, London.

Ceccato, V., Haining, R. and Signoretta, P. (2002). Exploring offence statistics in Stockholm City using spatial analysis tools. *Annals of the Association of American Geographers*, 92, 29–51.

Cecchini, A. (1996). Urban modeling by means of cellular automata: generalized urban automata with the help on-line (AUGH) model. *Environment and Planning B*, 23, 721–732.

Cervero, R. (1991). Land use and travel at suburban activity centers. *Transportation Quarterly*, 45(4), 479–491.

Cervero, R. (2001). Efficient urbanization: economic performance and the shape of the metropolis. *Urban Studies*, 28, 1651–1671.

Chan, J.C.W., Chan, K.P. and Yeh, A.G.O. (2001). Detecting the nature of change in an urban environment: a comparison of machine learning algorithms. *Photogrammetric Engineering and Remote Sensing*, 67, 213–225.

Chen, J., Gong, P., He, C., Luo, W., Tamura, M. and Shi, P. (2002). Assessment of the urban development plan by using a CA-based urban growth model. *Photogrammetric Engineering and Remote Sensing*, 68(10), 1063–1071.

Chen, J., Gong, P., He, C., Pu, R. and Shi, P. (2003a). Land-use/land-cover change detection using improved change vector analysis. *Photogrammetric Engineering and Remote Sensing*, 69, 369–379.

Chen, K.S., Yen, S.K. and Tsay, D.W. (1997). Neural classification of SPOT imagery through integration of intensity and fractal information. *International Journal of Remote Sensing*, 18(4), 763–783.

Chen, Z., Chen, J., Shi, P. and Tamura, M. (2003b). An IHS-based change detection approach for assessment of urban expansion impact on arable land loss in China. *International Journal of Remote Sensing*, 24, 1353–1360.

Cheng, J. and Masser, I. (2003). Urban growth pattern modeling: a case study of Wuhan city, PR China. *Landscape Urban Plan*, 62, 199–217.

Cho, S. and Yen, S.T. (2007). The differentiated impacts of urban growth boundary on land value and development between urban and rural–urban interface areas. *Lincoln Institute of Land Policy Paper Series*, Lincoln Institute.

Cho, S., Chen, Z., Yen, S.T. and Eastwood, D.B. (2006). Estimating effects of an urban growth boundary on land development. *Journal of Agricultural and Applied Economics*, 38, 287–298.

Civco, D.L., Hurd, J.D., Wilson, E.H., Arnold, C.L. and Prisloe, S. (2002). Quantifying and describing urbanizing landscapes in the Northeast United States. *Photogrammetric Engineering and Remote Sensing*, 68(10), 1083–1090.

Clapham, W.B., Jr. (2003). Continuum-based classification of remotely sensed imagery to describe urban sprawl on the watershed scale. *Remote Sensing of Environment*, 86, 322–340.

Clark, D. (1982). *Urban Geography: An Introductory Guide*. Taylor & Francis, London, p. 231.

Clarke, K. C., Hoppen, S. and Gaydos, L. (1997). A self-modifying cellular automaton model of historical urbanization in the San Francisco Bay area. *Environment and Planning B*, 24, 247–261.

Clarke, K.C. and Gaydos, L.J. (1998). Loose-coupling a cellular automaton model and GIS: long-term urban growth prediction for San Francisco and Washington/Baltimore. *International Journal of Geographical Information Science*, 12, 699–714.

Clarke, K.C. and Hoppen, S. (1997). A self-modifying cellular automaton model of historical urbanization in the San Francisco Bay area. *Environment and Planning B*, **24**, 247–261.

Clarke, K.C., Hoppen, S. and Gaydos, L.J. (1996). Methods and techniques for rigorous calibration of a cellular automaton model of urban growth. In *Proceedings of 3rd International Conference/Workshop on Integrating GIS and Environmental Modeling*, Santa Fe NM, January 21–26, National Center for Geographic Information and Analysis, Santa Barbara CA. URL: http://www.ncgia.ucsb.edu/conf/SANTA_FE_CD-ROM/main.html.

Clarke, K.C., Parks, B.O. and Crane, M.P. (2002). *Geographic Information Systems and Environmental Modeling*. Prentice Hall, Upper Saddle River, NJ.

Clawson, M. (1962). Urban sprawl and speculation in suburban land. *Land Economics*, **38**(2), 99–111.

Cliff, A. and Ord. J. (1975). Model building and the analysis of spatial pattern in human geography. *Journal of the Royal Statistical Society B*, **37**, 297–348.

Conte, C.R. (2000). The boys of sprawl. *Governing*, May Issue, 28–33.

Conzen, M.P. (2001). The study of urban form in the United States. *Urban Morphology*, **5**(1), 3–14.

Cooley, T.F. and La Civita, C.J. (1982). A theory of growth controls. *Journal of Urban Economics*, **12**, 129–145.

Coppin, P., Jonckheere, I., Nackaerts, K., Muys, B. and Lambin, E. (2004). Digital change detection methods in ecosystem monitoring: a review. *International Journal of Remote Sensing*, **25**(9), 1565–1596.

Coppin, P.R. and Bauer, M.E. (1994). Processing of multitemporal Landsat TM imagery to optimize extraction of forest cover change features. *IEEE Transactions on Geoscience and Remote Sensing*, **32**, 918–927.

Corne, S., Murray, T., Openshaw, S., See, L. and Turton, I. (1999). Using computational intelligence techniques to model sub glacial water systems. *Journal of Geographical System*, **1**, 37–60.

Couch, C. and Karecha, J. (2006). Controlling urban sprawl: some experiences from Liverpool. *Cities*, **23**(5), 353–363.

Couclelis, H. (2002). Modeling frameworks, paradigms, and approaches. In: K.C. Clarke, B.E. Parks and M.P. Crane (eds.), *Geographic Information Systems and Environmental Modeling*, Prentice Hall, London.

Coughlin, R. (1991). Formulating and evaluating agricultural zoning programs, *Journal of the American Planning Association*, **57**(2), 183–192.

Craglia M., Haining R. and Signoretta P. (2001). Modelling high-intensity crime areas in English cities. *Urban Studies*, **38**, 1921–1941.

Craglia M., Haining R. and Wiles P. (2000). A comparative evaluation of approaches to urban crime pattern analysis. *Urban Studies*, **37**, 711–729.

Crawford, T.W. (2007). Where does the coast sprawl the most? Trajectories of residential development and sprawl in coastal North Carolina, 1971–2000. *Landscape and Urban Planning*, **83**, 294–307.

Curran, P.J. (1987). Remote sensing methodologies and geography. *International Journal of Remote Sensing*, **8**, 1255–1275.

Dai, X. and Khorram, S. (1999). Remotely sensed change detection based on artificial neural networks. *Photogrammetric Engineering and Remote Sensing*, **65**, 1187–1194.

Danielsen, K.A., Lang, R.E. and Fulton, W. (1999). Retracting suburbia: Smart growth and the future of housing. *Housing Policy Debate*, **10**, 513–540.

Davis, C. and Schaub, T. (2005). A transboundary study of urban sprawl in the Pacific Coast region of North America: The benefits of multiple measurement methods. *International Journal of Applied Earth Observation and Geoinformation*, **7**, 268–283.

De la Barra, T. (1989). *Integrated Land Use and Transport Modelling*. Cambridge University Press, Cambridge.

De Roo, G. and Miller, D. (2000). *Compact Cities and Sustainable Urban Development: A Critical Assessment of Policies and Plans from an International Perspective*. Ashgate, Hampshire.

References

Deadman, P., Brown, R.D. and Gimblett, P. (1993). Modelling rural residential settlement patterns with cellular automata. *Journal of Environment Management*, 37, 147–160.

Deal, B. and Schunk, D. (2004). Spatial dynamic modeling and urban land use transformation: a simulation approach to assessing the costs of urban sprawl. *Ecological Economics*, 51, 79–95.

DeGrove, J.M. and Deborah, A.M. (1992). *The New Frontier for Land Policy: Planning and Growth Management in the States*. Lincoln Institute of Land Policy, Cambridge, MA.

Dendrinos, D.S. and Mullally, H. (1985). *Urban Evolution: Studies in the Mathematical Ecology of Cities*. Oxford University Press, Oxford.

Dewan, A.M. and Yamaguchi, Y. (2009). Land use and land cover change in greater Dhaka, Bangladesh: using remote sensing to promote sustainable urbanization. *Applied Geography*, 29, 390–401.

Diappi, L., Bolchim, P. and Buscema, M. (2004). Improved understanding of urban sprawl using neural networks. In: J.P. van Leeuwen and H.J.P. Timmermans (eds.), *Recent Advances in Design and Decision Support Systems in Architecture and Urban Planning*, Kluwer Academic Publishers, Dordrecht, pp. 33–49.

Dietzel, C., Herold, M., Hemphill, J.J. and Clarke, K.C. (2005). Spatio-temporal dynamics in California's Central Valley: empirical links urban theory. *International Journal of Geographic Information Sciences*, 19(2), 175–195.

DiLorenzo, W. (2000). The myth of suburban sprawl. *USA Today*, 128(May), 54–56.

Ding, C. (1998). The GIS-based human-interactive TAZ design algorithm: examining the impacts of data aggregation on transportation-planning analysis. *Environment and Planning B*, 25, 601–616.

Ding, C., Knaap G.J. and Hopkins, L.D. (1999). Managing urban growth with urban growth boundaries: a theoretical analysis. *Journal of Urban Economics*, 46, 530–568.

Dong, Y., Forster, B. and Ticehurst, C. (1997). Radar backscatter analysis for urban environments. *International Journal of Remote Sensing*, 18(6), 1351–1364.

Donnay, J.P., Barnsley, M.J. and Longley, P.A. (2001). Remote sensing and urban analysis. In: J.P. Donnay, M.J. Barnsley and P.A. Longley (eds.), *Remote Sensing and Urban Analysis*, Taylor & Francis, London and New York, pp 3–18.

Downing, P.B. and Gustely R.D. (1977). The public service costs of alternative development patterns: a review of the evidence. In: P.B. Downing (ed.), *Local Service Pricing Policies and Their Effect on Urban Spatial Structure*, University of British Columbia Press, Vancouver, BC.

Downs, A. (1999). Some realities about sprawl and urban decline. *Housing Policy Debate*, 10(4), 955–974.

Downs, A. (2001). What does 'smart growth' really mean? *Planning*, 67, 20–25.

Downs, A. (2005). Smart growth: Why we discuss it more than we do it. *Journal of the American Planning Association*, 71, 467–488.

Doxani, G., Siachalou, S. and Tsakiri-Strati, M. (2008). An object-oriented approach to urban land cover change detection. *The International Archives of the Photogrammetry, Remote Sensing and Spatial Information Sciences,* 37(B7), 1655–1660.

Du, G. (2001). Using GIS for analysis of urban systems. *GeoJournal*, 52, 213–221.

Du, Y., Guindon, B. and Cihlar, J. (2002). Haze detection and removal in high resolution satellite images with wavelet analysis. *IEEE Transactions on Geoscience and Remote Sensing*, 40, 210–217.

Duany, A., Plater-Zyberk, E. and Speck, J. (2001). *Suburban Nation: The Rise of Sprawl and the Decline of the American Dream*. North Point Press, New York.

DuBose, P. and Klimasauskas, C. (1989). *Introduction to Neural Networks with Examples and Applications*. NeuralWare Inc., Pittsburgh.

Duda, R.O., Hart, P.E. and Stork, D.G. (2001). *Pattern Classification*. Wiley, New York.

Duggin, M.J., Rowntree, R., Emmons, M., Hubbard, N., Odell, A.W., Sakhavat, H. and Lindsay, J. (1986). The use of multidate multichannel radiance data in urban feature analysis. *Remote Sensing of the Environment*, 20, 95–105.

Duncan, C. and Jones, K. (2000). Using multilevel models to model heterogeneity: potential and pitfalls. *Geographical Analysis*, 32, 279–305.

Dunn, C.P., Sharpe, D.M., Guntensbergen, G.R., Stearns, F. and Yang, Z. (1991). Methods for analyzing temporal changes in landscape pattern. In: M.G. Turner and R.H. Gardner (eds.), *Quantitative Methods in Landscape Ecology: The Analysis and Interpretation of Landscape Heterogeneity*, Springer Verlag, New York, pp. 173–198.

Dwyer, J.F. and Childs, G.M. (2004). Movement of people across the landscape: a blurring of distinctions between areas, interests and issue affecting natural resource management. *Landscape and Urban Planning*, **69**, 153–164.

Easley, V.G. (1992). *Staying Inside the Line, Planning Advisory Service Report Number 440*, American Planning Association, Chicago, IL, pp. 2–6.

Eastman, J.R. and Fulk, M. (1993). Long sequence time series evaluation using standardized principle components. *Photogrammetric Engineering and Remote Sensing*, **59**(6), 991–996.

El Nasser, H. and Overberg, P. (2001). A comprehensive look at sprawl in America. *USA Today*, February 22.

Ellman, T. (1997). *Infill: The Cure for Sprawl? Arizona Issue Analysis, 146*. The Goldwater Institute, Phoenix, AZ, p. 21.

Engle, R., Navarro, P. and Carson, R. (1992). On the theory of growth controls. *Journal of Urban Economics*, **32**, 269–283.

English, M. (1999). A guide for smart growth. *Forum for Applied Research and Public Policy*, **14**(3), 35–39.

English, M.R. and Hoffman, R.J. (2001). *Planning for Rural Areas in Tennessee Under Public Chapter 1101*, White Paper Presented for the TACIR—The Tennessee Advisory Commission on Intergovernmental Relations. URL: http://www.state.tn.us/tacir/Portal/Reports/Rural%20ares.pdf.

EPA (Environmental Protection Agency) (2000). *Projecting Land Use Change: A Summary of Models for Assessing the Effects of Community Growth and Change on Land Use Pattern*. URL: http://www.epa.gov/cbep/tools/reportfinal3.pdf.

ERDAS (2008). *ERDAS Field Guide, vol. 1&2*. Leica Geosystems Geospatial Imaging, Norcross, GA.

Estes, J.E., Stow, D. and Jensen, J.R. (1982). Monitoring land use and land cover changes. In: C.J. Johannsen and J.L. Sanders (eds.), *Remote Sensing for Resource Management*. Soil Conservation Society of America, Ankeny, IA, pp. 100–110.

Ewing, R. (1991). *Developing Successful New Communities*. Urban Land Institute, Washington, DC.

Ewing, R. (1994). Characteristics, causes, and effects of sprawl: a literature review. *Environmental and Urban Studies*, **21**(2), 1–15.

Ewing, R. (1997). Is Los Angeles-style sprawl desirable? *Journal of the American Planning Association*, **63**(1), 107–126.

Ewing, R., Pendall, R. and Chen, D.D.T. (2002). *Measuring Sprawl and Its Impact*. Smart Growth America, Washington, DC.

Fagan, W.F., Meir, E., Carroll, S.S. and Wo, J. (2001). The ecology of urban landscape: modeling housing starts as a density-dependent colonization process. *Landscape Ecology*, **16**, 33–39.

Farina, A. (1998). *Principles and Methods in Landscape Ecology*. Chapman & Hall, London.

Farley, R. and Frey, W.H. (1994). Changes in the segregation of whites from blacks in the 1980s: small steps toward a more integrated society. *American Sociological Review*, **59**, 23–45.

Faust, N.L. (1989). Image Enhancement. In: A. Kent and J. G. Williams (eds), *Encyclopedia of Computer Science and Technology*, Vol. 20, Supl. 5, Marcel Dekker Inc., New York.

Fischel, W.A. (1982). The urbanization of agricultural land: a review of the National Agricultural Lands Study. *Land Economics*, **58**(2), 236–259.

Fischer, M.M. (1995). Fundamentals in neuro-computing. In: M.M. Fischer, T. Sikos and L. Bassa (eds.), *Recent Developments in Spatial Information, Modeling and Processing*, Geomarket, Budapest, pp. 31–43.

Fischer, M.M., Gopal, S., Staufer, P. and Steinnocher, K. (1997). Evaluation of neural pattern classifiers for a remote sensing application. *Geographical Systems*, **4**(2), 195–226.

Flint, C. (2002). The theoretical and methodological utility of space and spatial statistics for historical studies: the Nazi Party in geographic context. *Historical Methods*, **35**, 32–42.

Fodor, E. (1999). *Better Not Bigger: How to Take Control of Urban Growth and Improve Your Community*, New Society Publishers, Gabriola Island, BC, p. 175.

Foresman, T.W., Pickett, S.T.A. and Zipperer, W.C. (1997). Methods for spatial and temporal land use and land cover assessment for urban ecosystems and application in the greater Baltimore–Chesapeake region. *Urban Ecosystems*, **1**(4), 201–216.

Forman, R.T.T. (1995). *Land Mosaics: The Ecology of Landscapes and Regions*. Cambridge University Press, Cambridge.

Forman, R.T.T. and Godron, M. (1986). *Landscape Ecology*. John Wiley and Sons, New York.

Forrester, J.W. (1969). *Urban Dynamics*. MIT Press, Cambridge, MA.

Forster, B. (1983). Some urban measurements from Landsat data. *Photogrammetric Engineering and Remote Sensing*, **14**, 1693–1707.

Forster, B. (1985). An examination of some problems and solutions in monitoring urban areas from satellite platforms. *International Journal of Remote Sensing*, **6**, 139–151.

Fotheringham, A.S. (1989). Scale-independent spatial analysis. In: M. Goodchild and S. Gopal (eds.), *Accuracy of Spatial Databases*, Taylor & Francis, London, pp. 221–228.

Fotheringham, A.S. (2000). GIS-based spatial modelling: a step forwards or a step backwards. In: A.S. Fotheringham and M. Wegener (eds.), *Spatial Models and GIS: New Potential and New Models*, Taylor & Francis, London, pp. 21–30.

Fotheringham, A.S. and O'Kelly, M.E. (1989). *Spatial Interaction Models: Formulations and Applications*, Kluwer Academic, Dordrecht.

Fotheringham, A.S. and Wegener, M. (2000). *Spatial Models and GIS: New Potential and New Models*. Taylor & Francis, London, p. 279.

Fotheringham, A.S. and Wong D.W.S. (1991). The modifiable areal unit problem in multivariate statistical analysis. *Environment and Planning A*, **23**, 1025–1044.

Fotheringham, A.S., Brunsdon C. and Charlton M. (2000). *Quantitative Geography: Perspectives on Spatial Data Analysis*. Sage, London.

Fotheringham, A.S., Densham P.J. and Curtis A. (1995). The zone definition problem in location-allocation modeling. *Geographical Analysis* **27**, 60–77.

Frank, J.E. (1989). *The Costs of Alternative Development Patterns: A Review of the Literature*. Urban Land Institute, Washington, DC.

Frankhauser, P. (1994). *La Fractalité des Structures urbaines*, Anthropos, Paris.

Frenkel, A. (2004). The potential effect of national growth-management policy on urban sprawl and the depletion of open spaces and farmland, *Land Use Policy*, **21**, 357–369.

Frumkin, H. (2002). Urban sprawl and public health. *Public Health Reports*, **117**, 201–217.

Fujita, M., Krugman, P. and Venables, A.J. (1999). *The Spatial Economy: Cities, Regions, and International Trade*, MIT Press, Cambridge, MA.

Fulton, W., Pendall, R., Nguyen, M. and Harrison, A. (2001). *Who Sprawls the Most? How Growth Patterns Differ Across the US*. The Brookings Institution Center on Urban and Metropolitan Policy, Washington, DC.

Fung, T. (1990). An assessment of TM imagery for land-cover change detection. *IEEE Transactions on Geoscience and Remote Sensing*, **28**, 681–684.

Fung, T. (1992). Land use and land cover change detection with Landsat MSS and SPOT HRV data in Hong Kong. *Geocarto International*, **3**, 33–40.

Fung, T. and LeDrew, E. (1987). Application of principal components analysis change detection. *Photogrammetric Engineering and Remote Sensing*, **53**(12), 1649–1658.

Fung, T. and LeDrew, E. (1988). The determination of optimal threshold levels for change detection using various accuracy indices. *Photogrammetric Engineering and Remote Sensing*, **54**, 1449–1454.

Galster, G. (1991). Black suburbanization: has it changed the relative location of the races? *Urban Affairs Quarterly*, **26**, 621–628.

Galster, G., Hanson, R. and Wolman, H., et al. (2000). *Wrestling Sprawl to the Ground: Defining and Measuring an Elusive Concept*. Working Paper of Fannie Mae, pp. 1–38.

Galster, G., Hanson, R., Wolman, H., Coleman, S. and Freihage, J. (2001). Wrestling sprawl to the ground: defining and measuring an elusive concept. *Housing Policy Debate*, 12(4), 681–717.

Gamba, P. and Dell'Acqua, F. (2003). Increased accuracy multiband urban classification using a neuro-fuzzy classifier. *International Journal of Remote Sensing*, 24(4), 827–834.

Gamba, P. and Houshmand, B. (2001). An efficient neural classification of SAR and optical urban images. *International Journal of Remote Sensing*, 22, 1535–1553.

Gao, J. and Skillcorn, D. (1998). Capability of SPOT XS data in producing detailed land cover maps at the urban–rural periphery. *International Journal of Remote Sensing*, 19(15), 2877–2891.

Gatrell, J.D. and Jensen, R.R. (2002). Growth through greening: developing and assessing alternative economic development programmes. *Applied Geography*, 22, 331–350.

Gatrell, J.D. and Jensen, R.R. (2008). Sociospatial applications of remote sensing in urban environments. *Geography Compass*, 2(3), 728–743.

Gehlke, C.E. and Biehl, K. (1934). Certain effects of grouping upon the size of the correlation coefficient in census tract material. *Journal of the American Statistical Association*, 29, 169–170.

Geoghegan, J., Wainger, L.A. and Bockstael, N.E. (1997). Spatial landscape indices in a hedonic framework: an ecological economics analysis using GIS. *Ecological Economics*, 23(3), 251–264.

Getis, A. and Ord, J.K. (1993). The analysis of spatial association by use of distance statistics. *Geographical Analysis*, 25, 276–276.

Geymen, A. and Baz, I. (2008). Monitoring urban growth and detecting land-cover changes on the Istanbul metropolitan area. *Environmental Monitoring Assessment*, 136, 449–459.

Gilbert, R., Stevenson, D., Giradet, H. and Stren, R. (1996). *Making Cities Work: The Role of Local Authorities in the Urban Environment*, Earthscan, London.

Gillham, O. and Maclean, A. (2001). *The Limitless City*. Island Press, Washington, DC.

Gimblett, R.H. and Ball, G.L. (1995). Neural network architectures for monitoring and simulating changes in forest resource management. *AI Applications*, 9(2), 103–123.

Giudici, M. (2002). Development, calibration, and validation of physical models. In: K.C. Clarke, B.O. Parks and M.P. Crane (eds.), *Geographic Information Systems and Environmental Modeling*, Prentice Hall, Upper Saddle River, NJ.

Giuliano, G. (1989). *Literature Synthesis: Transportation and Urban Form*. Report prepared for the Federal Highway Administration under Contract DTFH61-89-P-00531.

Glaeser, E., Kahn, M. and Chu, C. (2001). Job sprawl: employment location in U.S. metropolitan areas. *Survey Series of Center for Urban & Metropolitan Policy*, The Brookings Institution, Washington, DC, pp. 1–8.

Glaeser, E.L. and Kahn, M.E. (2004). Sprawl and urban growth. In: V. Henderson and J. Thisse (eds.), *The Handbook of Urban and Regional Economics*, Oxford University Press, Oxford.

Golledge, R.G. (1995). Primitives of spatial knowledge. In: T.L. Nyerges, D.M. Mark, R. Laurini and M.J. Egenhofer (eds.), *Cognitive Aspects of Human-Computer Interaction for Geographic Information Systems*, Springer, New York, pp. 29–44.

Gomiero, T., Giampietro, M., Bukkens, S.G.F. and Paoletti, M.G. (1999). Environmental and socioeconomic constraints to the development of freshwater fish aquaculture in China. *Critical Reviews in Plant Sciences*, 18(3), 359–371.

Gong, P., Marceau, D.J. and Howarth, P.J. (1992). A comparison of spatial feature extraction algorithms for land-use classification with SPOT HRV data. *Remote Sensing of Environment*, 40, 137–151.

Gonzalez, R.C. and Wintz, P. (1977). *Digital Image Processing*. Addison-Wesley Publishing Company, Reading, MA.

Goodchild, M.F. (1979). Aggregation problem in location-allocation. *Geographical Analysis*, 11, 240–255.

Gordon, P. and Richardson, H.W. (2000). *Critiquing Sprawl's Critics*. Cato Institute of Policy Analysis No: 365, Washington, DC.

References

Gordon, P. and Richardson, H.W. (1997a). Are compact cities a desirable planning goal? *Journal of the American Planning Association*, **63**(1), 95–106.

Gordon, P. and Richardson, H.W. (1997b). Where's the sprawl? *Journal of the American Planning Association*, **63**(2), 275–278.

Gordon, P. and Wong, H.L. (1985). The costs of urban sprawl—some new evidence. *Environment and Planning A*, **17**(5), 661–666.

Gordon, P., Kumar, A. and Richardson, H.W. (1989). The influence of metropolitan spatial structure on commuting time. *Journal of Urban Economics*, **26**, 138–151.

Gower, S.T., Kucharik, C.J. and Norman, J.M. (1999). Direct and indirect estimation of leaf area index, f(APAR), and net primary production of terrestrial ecosystems. *Remote Sensing of Environment*, **70**(1), 29–51.

Graham, J.U. and Fingelton, B. (1985). *Spatial Data Analysis by Example Volume 1: Point Pattern and Quantitative Data*, John Wiley & Sons, New York.

Grey, W. and Luckman, A. (2003). Mapping urban extent using satellite radar interferometry. *Photogrammetric Engineering and Remote Sensing*, **69**, 957–962.

Griffith, D.A. (1988). *Advanced Spatial Statistics: Special Topics in the Exploration of Quantitative Spatial Data Series*. Kluwer, Dordrecht.

Grimm, N.B., Grove, J.M., Pickett, S.T.A. and Redman, C.L. (2000). Integrated approaches to long-term studies of urban ecological systems. *BioScience*, **50**, 571–584.

Grove, J.M. and Burch, W.R., Jr. (1997). A social ecology approach and applications of urban ecosystem and landscape analyses: a case study of Baltimore, Maryland. *Urban Ecosystems*, **1**, 259–275.

Guo, H., Liu, L., Huang, G., Fuller, G., Zou, R. and Yin, Y. (2001). A system dynamics approach for regional environmental planning and management: a study for the Lake Erhai Basin. *Journal of Environmental Management*, **61**, 93–111.

Guo, J.Y. and Bhat, C.R. (2004). Modifiable areal units: Problem or perception in modeling of residential location choice? *Transportation Research Record*, **1898**, 138–147.

Gustafson, E.J. (1998). Quantifying landscape spatial pattern: what is the state of the art? *Ecosystems*, **1**, 143–156.

Haack, B., Bryant, N. and Adams, S. (1987). An assessment of Landsat MSS and TM data for urban and near-urban land–cover digital classification. *Remote Sensing of the Environment*, **20**, 201–213.

Hadly, C.C. (2000). *Urban Sprawl: Indicators, Causes, and Solutions*. Document prepared for the Bloomington Environmental Commission. URL: http://www.city.bloomington.in.us/planning.

Haggett, P., Cliff, A. and Frey. A. (1977). Locational methods. Halsted Press, London.

Haider, M. and Miller, E.J. (2000). Effects of Transportation Infrastructure and Location on Residential Real Estate Values: Application of Spatial Autoregressive Techniques. *Transportation Research Record*, **1722**, 1–8. URL: http://dx.doi.org/10.3141/1722-01.

Haines, V. (1986). Energy and urban form: a human ecological critique. *Urban Affairs Quarterly*, **21**(3), 337–353.

Hanham, R. and Spiker, J.S. (2007). Urban sprawl detection using satellite imagery and geographically weighted regression. In: *Geo-Spatial Technologies in Urban Environments Policy, Practice, and Pixels*, 2nd ed., Springer, Berlin, Heidelberg, doi:10.1007/978-3-540-69417-5, pp. 137–151.

Hanham, R., Hoch, R.J. and Spiker, J.S. (2009). The spatially varying relationship between local land-use policies and urban growth: a geographically weighted regression analysis. In: J.D. Gatrell and R.R. Jensen (eds.), *Planning and Socioeconomic Applications*, Springer, Berlin, pp. 43–56.

Hannah, L., Kim, K.H. and Mills, E.S. (1993). Land use controls and housing prices in Korea. *Urban Studies*, **30**, 147–156.

Hardin, P.J. (2000). Neural networks versus nonparametric neighbor-based classifiers for semisupervised classification of Landsat Thematic Mapper imagery. *Optical Engineering*, **39**, 1898–1908.

Hardin, P.J., Jackson, M.W. and Otterstrom, S.M. (2007). Mapping, measuring, and modeling urban growth. In: R.R. Jensen, J.D. Gatrell and D. McLean (eds.), *Geo-Spatial Technologies in Urban Environments*, 2nd ed., Springer, Berlin, pp. 141–176.

Hargis, C.D., Bissonette, J.A. and David, J.L. (1998). The behavior of landscape metrics commonly used in the study of habitat fragmentation. *Landscape Ecology*, 13, 167–186.

Harris, B. (1965). Urban development models: a new tool for planners. *Journal of the American Institute of Planners*, 31, 90–95.

Harris B. (1985). Urban simulation-models in regional science. *Journal of Regional Science*, 25, 545–567.

Harris, C.D. and Ullman, E.L. (1945). The nature of cities. *Annals of the American Academy of Political Science*, 242, 7–17.

Harris, J.R., Murray, R. and Hirose, T. (1990). IHS transform for the integration of radar imagery and other remotely sensed data. *Photogrammetric Engineering and Remote Sensing*, 56, 1631–1641.

Harvey, R.O. and Clark, W.A.V. (1965). The nature and economics of urban sprawl. *Land Economics*, 41(1), 1–9.

Hasna, A.M. (2007). Dimensions of sustainability. *Journal of Engineering for Sustainable Development: Energy, Environment, and Health*, 2(1), 47–57.

Hasse, J. (2004). A geospatial approach to measuring new development tracts for characteristics of sprawl. *Landscape Journal*, 23(1), 52–67.

Hasse, J. and Lathrop, R.G. (2003a). A housing-unit-level approach to characterizing residential sprawl. *Photogrammetric Engineering and Remote Sensing*, 69(9), 1021–1030.

Hasse, J.E. and Lathrop, R.G. (2003b). Land resource impact indicators of urban sprawl. *Applied Geography*, 23, 159–175.

Hathout, S. (2002). The use of GIS for monitoring and predicting urban growth in East and West St Paul, Winnipeg, Manitoba, Canada. *Journal of Environmental Management*, 66, 229–238.

Haykin, S. (1994). *Neural Networks: a Comprehensive Foundation*. Prentice-Hall, Upper Saddle River, NJ.

HCPC (Howard County Planning Commission) (1967). *Howard County 1985*. HCPC, Howard County, MD.

He, C., Okada, N., Zhang, Q., Shi, P. and Zhang, J. (2006). Modeling urban expansion scenarios by coupling cellular automata model and system dynamic model in Beijing, China. *Applied Geography*, 26, 323–345.

Hedblom, M. and Soderstrom, B. (2008). Woodlands across Swedish urban gradients: status, structure and management implications. *Landscape and Urban Planning*, 84, 62–73.

Heimlich, R.E. and Anderson, W.D. (2001). *Development at the Urban Fringe and Beyond: Impacts on Agriculture and Rural Land*. ERS Agricultural Economic Report No. 803, p. 88.

Helsley, R.W. and Strange, W.C. (1995). Strategic growth controls. *Regional Science and Urban Economics*, 25, 435–460.

Hepner, G.F., Houshmand, B., Kulikov, I. and Bryant, N. (1998). Investigation of the integration of AVIRIS and IFSAR for urban analysis. *Photogrammetric Engineering and Remote Sensing*, 64, 813–820.

Herold, M., Clarke, K.C. and Scepan, J. (2002). Remote sensing and landscape metrics to describe structures and changes in urban landuse. *Environment and Planning A*, 34(8), 1443–1458.

Herold, M., Couclelis, H. and Clarke, K.C. (2005a). The role of spatial metrics in the analysis and modeling of urban change. *Computers, Environment, and Urban Systems*, 29, 339–369.

Herold, M., Goldstein, N. and Clarke, K.C. (2003a). The spatio-temporal form of urban growth: measurement, analysis and modeling. *Remote Sensing of Environment*, 86(3), 286–302.

Herold, M., Hemphill, J., Dietzel, C. and Clarke, K.C. (2005b). Remote sensing derived mapping to support urban growth theory. *Proceedings of the ISPRS joint conference 3rd International Symposium Remote Sensing and Data Fusion Over Urban Areas, and 5th International Symposium Remote Sensing of Urban Areas (URS 2005)*, March 14–16, Tempe, AZ, USA. URL: www.isprs.org/commission8/workshop_urban/herold_hemphill_etal.pdf.

References

Herold, M., Liu, X. and Clarke, K.C. (2003b). Spatial metrics and image texture for mapping urban land use. *Photogrammetric Engineering and Remote Sensing*, **69**(8), 991–1001.

Herold, M., Roberts, D., Gardner, M. and P. Dennison (2004). Spectrometry for urban area remote sensing—development and analysis of a spectral library from 350 to 2400 nm. *Remote Sensing of Environment*, **91**, 304–319.

Heynen, N. and Lindsey, G. (2003). Correlates of urban forest canopy: implications for local public works. *Public Works Management and Policy*, **8**, 33–47.

Hobbs, R.J. (1999). Clark Kent or Superman: where is the phone booth for landscape ecology? In: J.M. Klopatek and R.H. Gardner (eds.), *Landscape Ecological Analysis: Issues and Applications*. Springer, New York.

Hodgson, M.J., Shmulevitz, F. and Körkel, M. (1997). Aggregation error effects on the discrete-space p-median model: the case of Edmonton, Canada. *Canadian Geographer*, **41**, 415–428.

Hollis, L.E. and Fulton, W. (2002). *Open Space Protection—Conservation Meets Growth Management*. The Brookings Institution Center on Urban and Metropolitan Policy, Washington, DC. URL: http://www.brookings.edu/reports/2002/ 04metropolitanpolicy_hollis.aspx.

Holt, D., Steel, D.G., Tranmer, M. and Wrigley, N. (1996). Aggregation and ecological effects in geographically based data. *Geographical Analysis*, **28**, 244–261.

Homer, C., Huang, C., Yang, L., Wylie, B. and Coan, M. (2004). Development of a 2001 National Landcover Database for the United States. *Photogrammetric Engineering and Remote Sensing*, **70**(7), 829–840.

Horner, M.W. and Murray, A.T. (2002). Excess commuting and the modifiable areal unit problem. *Urban Studies*, **39**, 131–139.

Howard, E. (1898). *Garden Cities of Tomorrow*. MIT Press, Cambridge, MA (London, 1902. Reprinted, edited with a Preface by F.J. Osborn and an Introductory Essay by Lewis Mumford).

Howarth, P.J. (1986). Landsat digital enhancements for change detection in urban environment. *Remote Sensing of Environment*, **13**, 149–160.

Howarth, P.J. and Wickware, G.M. (1981). Procedures for change detection using Landsat digital data. *International Journal of Remote Sensing*, **2**(3), 277–291.

Howell-Moroney, M. (2004). Community characteristics, open space preservation and regionalism: is there a connection? *Journal of Urban Affairs*, **26**, 109–118.

Hoyt, H. (1939). *The Structure and Growth of Residential Neighborhoods in American Cities*. Federal Housing Administration, Washington, DC.

Huang, J., Lu, X.X. and Sellers, J.M. (2007). A global comparative analysis of urban form: applying spatial metrics and remote sensing. *Landscape and Urban Planning*, **82**, 184–197.

HUD (1999). *The State of the Cities*. US Departmet of Housing and Urban Development, Washington, DC.

ICLEI (International Council for Local Environmental Initiatives) (2004). *Aalborg Commitments*, ICLEI. URL: http://www.aalborgplus10.dk/.

Ilachinski, A. (2001). *Cellular Automata: A discrete universe*. World Scientific, P. 808.

Im, J. and Jensen, J.R. (2005). A change detection model based on neighborhood correlation image analysis and decision tree classification. *Remote Sensing of Environment*, **99**, 326–340.

Imhoff, M.L., Lawrence, W.T., Stutzer, D.C. and Elvidge, C.D. (1997). A technique for using composite DMSP.OLS "city lights" satellite data to map urban area. *Remote Sensing of Environment*, **61**, 361–370.

Ingram, K., Knapp, E. and Robinson, J. (1981). *Change Detection Technique Development for Improved Urbanized Area Delineation*. Technical memorandum CSC/TM-81/6087 – Computer Sciences Corporation, Silver Springs, Maryland.

Irwin, E.G. and Bockstael, N.E. (2004). Land use externalities, open space preservation, and urban sprawl. *Regional Science and Urban Economics*, **34**, 705–725.

Irwin, E.G. and Geoghegan, J. (2001). Theory, data, methods: developing spatially explicit economic models of land use change. *Agriculture, Ecosystems and Environment*, **85**, 7–23.

Itami, R.M. (1994). Simulating spatial dynamics: cellular automata theory. *Landscape and Urban Planning*, **30**, 27–47.

IUCN (2006). *The Future of Sustainability: Re-thinking Environment and Development in the Twenty-first Century*. Report of the IUCN Renowned Thinkers Meeting, 29–31 January, International Union for Conservation of Nature. URL: http://cmsdata.iucn.org/ downloads/iucn_future_of_sustanability.pdf.

Jacobs, J. (1965). *The Death and Life of Great American Cities*, Penguin, London.

Jacquin, A., Misakova, L. and Gay, M. (2008). A hybrid object-based classification approach for mapping urban sprawl in periurban environment. *Landscape and Urban Planning*, 84, 152–165.

Jantz, C.A., Goetz, S.J. and Shelley, M.K. (2003). Using the SLEUTH urban growth model to simulate the impacts of future policy scenarios on urban land use in the Baltimore-Washington metropolitan area. *Environment and Planning B*, 30, 251–271.

Jat, M.K., Garg, P.K. and Khare, D. (2008). Modeling urban growth using spatial analysis techniques: a case study of Ajmer city (India). *International Journal of Remote Sensing*, 29(2), 543–567.

Jenerette, G.D. and Wu, J. (2001). Analysis and simulation of land-use change in the central Arizona Phoenix region, USA. *Landscape Ecology*, 16, 611–626.

Jensen, J.R. (2005). *Introductory Digital Image Processing: A Remote Sensing Perspective*. Prentice-Hall, Upper Saddle River, NJ.

Jensen, J.R. (2006). *Remote Sensing of the Environment: An Earth Resource Perspective*, 2nd ed., Prentice Hall, Upper Saddle River, NJ, p. 592.

Jensen, J.R. and Cowen, D.C. (1999). Remote sensing of urban/suburban infrastructure and socioeconomic attributes. *Photogrammetric Engineering and Remote Sensing*, 65(5), 611–622.

Jensen, J.R. and Im, J. (2007). Remote sensing change detection in urban environments. In: R.R. Jensen, J.D. Gatrell and D. McLean (eds.), *Geo-Spatial Technologies in Urban Environments: Policy, Practice and Pixels*, 2nd ed., Springer-Verlag, Heidelberg, pp. 7–30.

Jensen, J.R., Cowen, D.J., Althausen, J.D., Nanunalani, S. and Weatherbee, O. (1993). An evaluation of the CoastWatch change detection protocol in South Carolina. *Photogrammetric Engineering and Remote Sensing*, 59(6), 1039–1046.

Jensen, J.R., Hodgson, M.E., Tullis, J.A. and Raber, G.T. (2004a). Remote sensing of impervious surfaces and building infrastructure. In: R.R. Jensen, J.D. Gatrell and D. McLean (eds.), *Geospatial Technologies in Urban Environments: Policy, Practice and Pixels*, Springer, Heidelberg, pp. 5–20.

Jensen, J.R., Qui, F. and Ji, M. (2000). Predictive modeling of coniferous forest age using statistical and artificial neural network approaches applied to remote sensor data. *International Journal of Remote Sensing*, 20, 2805–2822.

Jensen, J.R., Rutchey, K., Koch, M.S. and Narumalani, S. (1995). Inland wetland change detection in the Everglades water conservation area 2A using a time series of normalized remotely sensed data. *Photogrammetric Engineering and Remote Sensing*, 61(2), 199–209.

Jensen, R., Gatrell, J., Boulton, J. and Harper, B. (2004b). Using remote sensing and geographic information systems to study urban quality of life and urban forest amenities. *Ecology & Society*, 9, 1–5.

Jensen, R.R. and Hardin, P.J. (2005). Estimating urban leaf area using field measurements and satellite remote sensing data. *GIScience and Remote Sensing*, 42, 229–252.

Jerrett, M., Burnett, R.T., Goldberg, M.S., Sears, M., Krewski, D., Catalan, R., Kanaroglou, P., Giovis, C. and Finkelstein, N. (2003). Spatial analysis for environmental health research: concepts, methods, and examples. *Journal of Toxicology and Environmental Health A*, 66, 1783–1810.

Ji, W., Ma, J., Twibell, R.W. and Underhill, K. (2006). Characterizing urban sprawl using multi-stage remote sensing images and landscape metrics. *Computers, Environment and Urban Systems*, 30, 861–879.

Jiang, F., Liu, S., Yuan, H. and Zhang, Q. (2007). Measuring urban sprawl in Beijing with geospatial indices. *Journal of Geographical Sciences*. doi:10.1007/s11442-007-0469-z.

Johnson, L.F. (2001a). Nitrogen influence on fresh-leaf NIR spectra. *Remote Sensing of Environment*, 78(3), 314–320.

References

Johnson, M.P. (2001b). Environmental impacts of urban sprawl: a survey of the literature and proposed research agenda. *Environment and Planning A*, **33**, 717–735.

Johnson, R.D. and Kasischke, E.S. (1998). Change vector analysis: a technique for the multi-temporal monitoring of land cover and condition. *International Journal of Remote Sensing*, **19**, 411–426.

Jones, K. (1991). Specifying and estimating multilevel models for geographical research. *Transactions of the Institute of British Geographers*, **16**, 148–159.

Juergensmeyer, J.C. (1984–1985). Implementing agricultural preservation programs: a time to consider some radical approaches? *Gonzaga Law Review*, **20**(3), 701–727.

Jun, M. (2004). The effects of Portland's urban growth boundary on urban development patterns and commuting. *Urban Studies*, **41**, 1333–1348.

Kahn, M. (2001). Does sprawl reduce black/white housing consumption gap? *Housing Policy Debate*, **12**, 77–86.

Kahn, M.E. and Schwartz, J. (2008). Urban air pollution progress despite sprawl: the "greening" of the vehicle fleet. *Journal of Urban Economics*, **63**(3), 775–787.

Kanaroglou, P.S. and Scott, D.M. (2002). Integrated urban transportation and land-use models. In: M. Dijst, W. Schenkel and I. Thomas (eds.), *Governing Cities on the Move: Functional and Management Perspectives on Transformations of European Urban Infrastructures*, Ashgate, Aldershot, pp. 42–72.

Kanaroglou, P.S., Soulakellis, N.A. and Sifakis, N.I. (2002). Improvement of satellite derived pollutions maps with the use of a geostatistical interpolation method. *Journal of Geographical Systems*, **4**, 193–208.

Kashian, R. and Skidmore, M. (2002). Preserving agricultural land via property assessment policy and the willingness to pay for land preservation. *Economic Development Quarterly*, **16**, 75–87.

Kato, S. and Yamaguchi, Y. (2005). Analysis of urban heat-island effect using ASTER and ETMþ data: separation of anthropogenic heat discharge and natural heat radiation from sensible heat flux. *Remote Sensing of Environment*, **99**(1&2), 44–54.

Kaufmann, R.K. and Seto, K.C. (2001). Change detection, accuracy, and bias in a sequential analysis of Landsat imagery in the Pearl River Delta, China: econometric techniques. *Agriculture, Ecosystems, and Environment*, **85**, 95–105.

Kawata, Y., Ohtani, A., Kusaka, T. and Ueno, S. (1990). Classification accuracy for the MOS-1 MESSR data before and after the atmospheric correction. *IEEE Transactions on Geoscience and Remote Sensing*, **28**, 755–760.

Kerridge, J., Hine, J. and Wigan, M. (2001). Agent-based modeling of pedestrian movements: the questions that need to be asked and answered. *Environment and Planning B*, **28**, 327–341.

Kim, J., Kang, Y., Hong, S. and Park, S. (2006). Extraction of spatial rules using a decision tree method: a case study in urban growth modeling. In: B. Gabrys, R.J. Howlett and L.C. Jain (eds.), *KES 2006, Part I, LNAI 4251*, Springer, Berlin, pp. 203–211.

King, R.B. (1994). The value of ground resolution, spectral range and stereoscopy of satellite imagery for land system and land-use mapping of the humid tropics. *International Journal of Remote Sensing*, **15**(3), 521–530.

Kirtland, D., Gaydos, L., Clarke, K., DeCola, L., Acevedo, W. and Bell, C. (1994). An analysis of human-induced land transformations in the San Francisco Bay/Sacramento area. *World Resources Review*, **6**(2), 206–217.

Kline, J.D. (2000). Comparing states with and without growth management analysis based on indicators with policy implications comment. *Land Use Policy*, **17**, 349–355.

Kline, J.D. (2005). Forest and farmland conservation effects of Oregon's (USA) land-use planning program. *Environmental Management*, **35**(4), 368–380.

Kline, J.D. (2006). Public demand for preserving local open space. *Society and Natural Resources*, **19**, 645–659.

Kline, J.D. and Alig, R.J. (1999). Does land-use planning slow the conversion of forest and farmlands, *Growth and Change*, **30**, 3–22.

Klinenberg, E. (2002). *Heat Wave: A Social Autopsy of Disaster in Chicago*. University of Chicago Press, Chicago.

Klosterman, R.E. (1999). The what if? Collaborative planning support system. *Environment and Planning B*, **26**, 393–408.

Knox, P.L. (1994). *Urbanization: introduction to urban Geography*. Prentice Hall, Upper Saddle River, p. 436.

Kohonen, T. (1995). *Self-organizing Maps*. Springer, Berlin.

Kropp, J. (1998). A neural network approach to the analysis of city systems. *Applied Geography*, **18**(1), 83–96.

Krummel, J.R., Gardner, R.H., Sugihara, G., O'Neill, R.V. and Coleman, P.R. (1987). Landscape patterns in a disturbed environment. *Oikos*, **48**, 321–324.

Kuby, M., Barranda, A. and Upchurch, C. (2004). Factors influencing light-rail station boardings in the United States. *Transportation Research Part A: Policy and Practice*, **38**, 223–247.

Kumar, J.A.V., Pathan, S.K. and Bhanderi, R.J. (2007). Spatio-temporal analysis for monitoring urban growth – a case study of Indore city. *Journal of Indian Society of Remote Sensing*, **35**(1), 11–20.

Kunstler, J.H. (1993). *The Geography of Nowhere*. Touchstone Books, New York.

Kunstler, J.H. (1998). *Home from Nowhere: Remaking Our Everyday World for the 21st Century*. Touchstone–Simon & Schuster, New York.

Kwan, M.P. (2000). Interactive geovisualization of activity-travel patterns using three-dimensional geographical information systems: a methodological exploration with a large data set. *Transportation Research C*, **8**, 185–203.

Lai, F. and Yang, S. (2002). A view on optimal urban growth controls. *Annals of Regional Science*, **36**, 229–238.

Lam, N.S.N. and Lee, D.C. (1993). *Fractals in Geography*. Prentice Hall, Upper Saddle River, p. 308.

Lambin, E.F., Turner, B.L., Geist, H.J., Agbola, S.B., Angelsen, A., Bruce, J.W., et al. (2001). The causes of land-use and land-cover change: moving beyond the myths. *Global Environmental Change*, **11**, 261–269.

Landis, J. and Zhang, M. (1998). The second generation of the California urban futures model Parts 1, 2 & 3. *Environment and Planning B*, **25**, 657–666 & 795–824.

Landis, J. and Zhang, M. (2000). Using GIS to improve urban activity and forecasting models: three examples. In: A.S. Fotheringham and M. Wegener (eds.), *Spatial Models and GIS: New Potential and New Models*, Taylor and Francis, London, pp. 63–81.

Lang, R.E. (2003). Open spaces, bounded places: does the American West's arid landscape yield dense metropolitan growth? *Housing Policy Debate*, **13**(4), 755–778.

Langford, M. and Bell, W. (1997). Land cover mapping in a tropical hillsides environment: a case study in the Cauca Region of Colombia. *International Journal of Remote Sensing*, **18**(6), 1289–1306.

Lassila, K.D. (1999). The new suburbanites: how American plants and animals are threatened by the sprawl. *The Amicus Journal*, **21**, 16–22.

Lata, K.M., Rao, C.H.S., Prasad, V.K., Badarianth, K.V.S. and Rahgavasamy, V. (2001). Measuring urban sprawl: a case study of Hyderabad. *GISdevelopment*, **5**(12), 26–29.

Ledermann, R.C. (1967). The city as a place to live. In: J. Gottmann and R.A. Harpe (eds.), *Metropolis on the Move: Geographers Look at Urban Sprawl*, Wiley, New York.

Lessinger, J. (1962). The cause for scatteration: some reflections on the National Capitol Region plan for the year 2000. *Journal of the American Institute of Planners*, **28**(3), 159–170.

Levine, J. and Landis, J. (1989). Geographic information systems for local planning. *Journal of the American Planning Association*, **55**, 209–220.

Levine, N. (1999). The effects of local growth controls on regional housing production and population redistribution in California. *Urban Studies*, **36**, 2047–2068.

Lewis, S. (1990). The town that said no to sprawl. *Planning (APA)*, **56**, 14–19.

Li, D.H. (1991). *Principle of Urban Planning*. China Architecture Industry Press, Beijing.

Li, X. and Yeh, A.G.O. (1998). Principal component analysis of stacked multi-temporal images for monitoring of rapid urban expansion in the Pearl River Delta. *International Journal of Remote Sensing*, **19**(8), 1501–1518.

References

Li, X. and Yeh, A.G.O. (2001). Calibration of cellular automata by using neural networks for the simulation of complex urban systems. *Environment and Planning A*, **33**, 1445–1462.

Li, X. and Yeh, A.G.O. (2004). Analyzing spatial restructuring of landuse patterns in a fast growing region remote sensing and GIS. *Landscape and Urban Planning*, **69**, 335–354.

Lindstrom, M.J. and Bartling, H. (2003). *Suburban Sprawl: Culture, Theory, and Politics*. Rowman and Littlefield, Lanham, MD.

Liu, J.G. and Mason, P.J. (2009). *Essential Image Processing and GIS for Remote Sensing*. Wiley-Blackwell, New York, p. 450.

Liu, S.X. and Zhu, X.A. (2004). Accessibility analyst: an integrated GIS tool for accessibility analysis in urban transportation planning. *Environment and Planning B*, **31**, 105–124.

Liu, X. (2000). Change detection for urban growth modeling: an artificial neural network approach. *Proceedings of 4th International Conference on Integrating GIS and Environmental Modeling (GIS/EM4): Problems, Prospects and Research Needs*, Banff, Alberta, Canada, September 2–8.

Liu, X. and Lathrop, R.G., Jr. (2002). Urban change detection based on an artificial neural network. *International Journal of Remote Sensing*, **23**, 2513–2518.

Liu, Y. and Phinn, S.R. (2003). Modeling urban development with cellular automata incorporating fuzzy-set approaches. *Computers, Environment and Urban Systems*, **27**(6), 637–658.

Liu, Y., Nishiyama, S. and Yano, T. (2004). Analysis of four change detection algorithms in bi-temporal space with a case study. *International Journal of Remote Sensing*, **25**, 2121–2139.

Liu, Y.S., Gao, J. and Yang, Y.F. (2003). A holistic approach towards assessment of severity of land degradation along the Greatwall in northern Shannxi province, China. *Environmental Monitoring and Assessment*, **82**, 187–202.

Lockwood, C. (1999). Sprawl. *Hemispheres*, September issue, 82–91.

Longley, P., Batty, M., Shepherd, J. and Sadler, G. (1992). Do green belts change the shape of urban areas? A preliminary analysis of the settlements geography of southeast England, *Regional Studies*, **26**(5), 437–452.

Longley, P.A. and Mesev, V. (2001). Measuring urban morphology using remotely-sensed imagery. In: J.P. Donnay, M.J. Barnsley and P.A. Longley (eds.), *Remote Sensing and Urban Analysis*, Taylor and Francis, London, pp. 163–183.

Longley, P.A. and Tobon, C. (2004). Spatial dependence and heterogeneity in patterns of hardship: an intra-urban analysis. *Annals of the Association of the American Geographers*, **94**(3), 503–519.

Longley, P.A. and Mesev, V. (2000). On the measurement and generalization of urban form. *Environment and Planning A*, **32**, 473–488.

Longley, P.A., Barnsley, M.J. and Donnay, J.P. (2001). Remote sensing and urban analysis: a research agenda. In: J.P. Donnay, M.J. Barnsley and P.A. Longley (eds.), *Remote Sensing and Urban Analysis*, Taylor and Francis, London and New York, pp. 245–258.

Lopez, E., Bocco, G., Mendoza, M. and Duhau, E. (2001). Predicting land cover and land use change in the urban fringe a case in Morelia City, Mexico. *Landscape and Urban Planning*, **55**(4), 271–285.

Lu, D. and Weng, Q. (2004). Spectral mixture analysis of the urban landscape in Indianapolis with Landsat ETM+ imagery. *Photogrammetric Engineering and Remote Sensing*, **70**(9), 1053–1062.

Lu, D. and Weng, Q. (2007). A survey of image classification methods and techniques for improving classification performance. *International Journal of Remote Sensing*, **28**(5), 823–870.

Lu, D., Mausel, P., Brondizios, E. and Moran, E. (2004). Change detection techniques. *International Journal of Remote Sensing*, **25**, 2365–2407.

Luck, M. and Wu, J. (2002). A gradient analysis of urban landscape pattern: a case study from the Phoenix metropolitan region, Arizona, USA. *Landscape Ecology*, **17**(4), 327–339.

MacDonald, K. and Rudel, T.K. (2005). Sprawl and forest cover: what is the relationship? *Applied Geography*, **25**, 67–79.

Macie, E. and Moll, G. (1989, July/August). Trees and exurban sprawl. *American Forests*, 61–64.
Madhavan, B.B., Kubo, S., Kurisaki, N. and Sivakumar, T.V.L.N. (2001). Appraising the anatomy and spatial growth of the Bangkok Metropolitan area using a vegetation-impervious-soil model through remote sensing. *International Journal of Remote Sensing*, **22**, 789–806.
Maktav, D. and Erbek, F.S. (2005). Analysis of urban growth using multitemporal satellite data in Istanbul, Turkey. *International Journal of Remote Sensing*, **26**, 797–810.
Maktav, D., Erbek, F.S. and Jürgens, C. (2005). Remote sensing of urban areas. *International Journal of Remote Sensing*, **26**(4), 655–659.
Malczewski, J. and Poetz, A. (2005). Residential burglaries and neighborhood socioeconomic context in London, Ontario: global and local regression analysis. *The Professional Geographer*, **57**, 516–529.
Malila, W.A. (1980). Change vector analysis: an approach for detecting forest changes with Landsat. *Proceedings of the 6th Annual Symposium on Machine Processing of Remotely Sensed Data*, June 3–6, Purdue University, West Lafayette, Indiana, pp. 326–335.
Malpezzi, S. (1999). *Estimates of the Measurement and Determinants of Urban Sprawl in U.S. Mteopolitan Areas*. Center for Urban Land Economics Research, University of Wisconsin, Madison, WI.
Mandelbrot, B.B. (1982). *The Fractal Geometry of Nature*. W.H. Freeman, San Francisco, p. 468.
Mandelbrot, B.B. (1983). *The Fractal Geometry of Nature*, 3rd edn. W.H. Freeman, San Francisco.
Manson, S. M. (2000). *Agent-Based Dynamic Spatial Simulation of Land-Use/Cover Change: Methodological Aspects*. University Consortium for Geographic Information Science Annual Meeting, Portland, Oregon. URL: http://www.ucgis.org/oregon/papers/manson.htm.
Martin, L.R.G. (1989). An evaluation of landsat-based change detection methods applied to the rural-urban fringe. In: C.R. Bryant, E.F. LeDrew, C. Marois and E. Cavayas (eds.), *Remote Sensing and Methodologies of Land Use Change Analysis*, University of Waterloo, Waterloo, Canada, pp. 101–116.
Martinuzzi, S., Gould, W.A. and Gonzalez, O.M.R. (2007). Land development, land use, and urban sprawl in Puerto Rico integrating remote sensing and population census data. *Landscape and Urban Planning*, **79**, 288–297.
Masek, J.G., Lindsay, F.E. and Goward, S.N. (2000). Dynamics of urban growth in the Washington DC metropolitan area, 1973–1996, from Landsat observations. *International Journal of Remote Sensing*, **21**, 3473–3486.
Mattson, G.A. (2002). *Small Towns, Sprawl and the Politics of Policy Choices: The Florida Experience*. University Press of America, Lanham, MD.
McArthur, R.H. and Wilson, E.O. (1967). *The Theory of Island Biogeography*. Princeton University Press, Princeton, NJ, p. 203.
Mcdonnell, M.J. and Pickett, S.T.A. (1990). Ecosystem structure and function along urban–rural gradients: an unexploited opportunity for ecology. *Ecology*, **71**, 1232–1237.
Mcdonnell, M.J., Pickett, S.T.A., Pouyat, R.V., Zipperer, W.C., Parmelee, R.W., Carreiro, M.M. and Medley, K. (1997). Ecosystem processes along an urban-to-rural gradient. *Urban Ecosystems*, **1**, 21–36.
McGarigal, K. and Marks, B.J. (1995). *FRAGSTATS: Spatial Pattern Analysis Program for Quantifying Landscape Structure*. USDA Forest Service General Technical Report PNW-351.
McGarigal, K., Cushman, S.A., Neel, M.C. and Ene., E. (2002). *FRAGSTATS: Spatial Pattern Analysis Program for Categorical Maps*. URL: www.umass.edu/landeco/research/fragstats/fragstats.html.
McHarg, I. (1969). *Design with Nature*. (reprinted on 1993), Wiley, New York.
McMillen, D.P. (2001). Nonparametric employment subcenter identification. *Journal of Urban Economics*, **50**, 448–473.
Meaille, R. and Wald, L. (1990). Using geographical information system and satellite imagery within a numerical simulation of regional growth. *International Journal of Geographical Information Systems*, **4**(4), 445–456.

References

Mennis, J. and Jordan, L. (2005). The distribution of environmental equity: exploring spatial non-stationarity in multivariate models of air toxic releases. *Annals of the Association of American Geographers*, **95**, 249–268.

Mesev, T.V., Longley, P.A., Batty, M. and Xie, Y. (1995). Morphology from imagery: detecting and measuring the density of urban land use. *Environment and Planning A*, **27**, 759–780.

Messina, J.P., Crews-Meyer, K.A. and Walsh, S.J. (2000). Scale dependent pattern metrics and panel data analysis as applied in a multiphase hybrid landcover classification scheme. *Proceedings of the 2000 ASPRS Conference*. URL: http://www.unc.edu/˜messina/asprs2000.html.

Meyer, M.D. and Miller, E.J. (2001). *Urban Transportation Planning*. 2nd Edition: McGraw-Hill, New York.

Miller, H.J. (2004). Tobler's first law and spatial analysis. *Annals of the Association of American Geographers*, **94**, 284–289.

Miller, H.J. (1999). Potential contributions of spatial analysis to geographic information systems for transportation (GIS-T). *Geographical Analysis*, **31**, 373–399.

Miller, H.J. and Shaw, S.L. (2001). *Geographic Information Systems for Transportation: Principles and Applications*. Oxford University Press, Oxford.

Miller, J.R. and Chen, J.M. (2001). Surface environment: earth observations. *La Physique Au Canada*, Septembre/Octobre, 287–296.

Mills, D.E. (1980). Growth, speculation, and sprawl in a monocentric city. *Journal of Urban Economics*, **10**, 201–226.

Millward, A.A., Piwowar, J.M. and Howarth, P.J. (2006). Time-series analysis of medium-resolution, multisensor satellite data for identifying landscape change. *Photogrammetric Engineering and Remote Sensing*, **72**, 653–663.

Milne, B.T. (1991). Lessons from applying fractal models to landscape patterns. In: M.G. Turner and R.H. Gardner (eds.), *Quantitative Methods in Landscape Ecology: The Analysis and Interpretation of Landscape Heterogeneity*, Springer-Verlag, New York, pp. 199–235.

Mitchell, J.G. (2001). Urban sprawl: the American dream? *National Geographic*, **200**(1), 48–73.

Moeckel, R., Spiekermann, K., Schürmann, C. and Wegener, M. (2003). Microsimulation of urban land use. *Proceedings of International Conference on Computers in Urban Planning and Urban Management*, Sendai, Japan, May 27–29.

Monmonier, M. (1996). *How to Lie with Maps* University of Chicago Press, Chicago.

Morisette, J.F. and Khorram, S. (2000). Accuracy assessment curves for satellite based change detection. *Photogrammetric Engineering and Remote Sensing*, **66**, 875–880.

MSP (Metropolitan Strategy Project) (2003). *Melbourne 2030*. Department of Infrastructure, Melbourne, pp. 1–10. URL: http://www.dse.vic.gov.au/melbourne2030online/downloads/plan1_ugb.pdf.

Mundia, C.N. and Aniya, M. (2005). Analysis of land use/cover changes and urban expansion in Nairobi city using remote sensing and GIS. *International Journal of Remote Sensing*, **26**, 2831–2849.

Murray, A. and Gottsegen, J. (1997). The influence of data aggregation on the stability of p-median location model solutions. *Geographical Analysis*, **29**, 200–213.

Najlis, R. and North, M.J. (2004). Repast for GIS. *Proceedings of Agent 2004: Social Dynamics: Interaction, Reflexivity and Emergence*, University of Chicago and Argonne National Laboratory, IL, USA.

NASA (2001). *Satellite Maps Provide Better Urban Sprawl Insight*. NASA News Release, 2 June 2001.

Neer, J.T. (1999). High resolution imaging from space—a commercial perspective on a changing landscape. *International Archives of Photogrammetry and Remote Sensing*, **32**(7C2), 132–143.

Nelson, A.C. (1990). Economic critique of prime farmland preservation policies in the United States. *Journal of Rural Studies*, **6**(2), 119–142.

Nelson, A.C. (1992). Preserving prime farmland in the face of urbanization: lessons from Oregon. *Journal of the American Planning Association*, **58**, 467–488.

Nelson, A.C. (1999). Comparing states with and without growth management: Analysis based on indicators with policy implications. *Land Use Policy*, **16**, 121–127.
Nelson, A.C. (2000). Comparing states with and without growth management analysis based on indicators with policy implications. *Land Use Policy*, **16**, 121–127.
Nelson, A.C. and Moore, T. (1993). Assessing urban growth management—the case of Portland, Oregon. *Land Use Policy*, **10**(4), 293–302.
Nelson, A.C., Pendall, R., Dawkins, C.J. and Knaap, G.J. (2002). *The Link Between Growth Management and Housing Affordability: The Academic Evidence*. The Brookings Institution Centre on Urban and Metropolitan Policy, Washington, DC.
Nelson, G.C., Nelson, D. and Hellerstein, D. (1997). Do roads cause deforestation? Using satellite images in econometric analysis of land-use. *American Journal of Agricultural Economics*, **79**, 80–88.
Nelson, R.E. (1983). Detecting forest canopy changes due to insect activity using Landsat MSS. *Photogrammetric Engineering and Remote Sensing*, **49**(9), 1303–1314.
Nelson, R.F. (1982). *Detecting Forest Canopy Change Using Landsat, NASA Technical Memorandum 83918*. Goddard Space Flight Centre, Greenbelt, MD.
Newman, P.W.G. and Kenworthy, J.R. (1988). The transport energy trade-off: fuel-efficient traffic versus fuel-efficient cities. *Transportation Research A*, **22A**(3), 163–174.
Noth, M., Borning, A. and Waddell, P. (2003). An extensible, modular architecture for simulating urban development, transportation, and environmental impacts. *Computers, Environment and Urban Systems*, **27**, 181–203.
NRC (National Research Council) (1991). *Spatial Statistics and Digital Image Analysis*. Panel on Spatial Statistics and Image Processing, Board on Mathematical Sciences, National Research Council, National Academics Press, USA, p. 246.
O'Connor, K.F., Overmars, E.B. and Ralston, M.M. (1990). *Land Evaluation for Nature Conservation*. Caxton Press, Wellington, New Zealand, p. 328.
O'Neill, R.V., Krummel, J.R., Gardner, R.H., Sugihara, G., Jackson, B., Deangelis, D.L., Milne, B.T., Turner, M.G., Zygmunt, B., Christensen, S.W., Dale, V.H. and Graham, R.L. (1988). Indices of landscape pattern. *Landscape Ecology*, **1**, 153–162.
O'Neill, R.V., Ritters, K.H., Wichham, J.D. and Jones, K.B. (1999). Landscape pattern metrics and regional assessment. *Ecosystem Health*, **5**, 225–233.
O'Sullivan, D. (2002). Toward micro-scale spatial modeling of gentrification. *Journal of Geographical Systems*, **4**, 251–274.
O'Sullivan, D. and Torrens, P.M. (2000). Cellular models of urban systems. In: S. Bandini and T. Worsch (eds.), *Theoretical and Practical Issues on Cellular Automata*, Springer, London, pp. 108–116.
Ohls, J.C. and Pines D. (1975). Discontinuous urban development and economic efficiency. *Land Economics*, **51**(3), 224–234.
Okabe, A. and Tagashira, N. (1996). Spatial aggregation bias in a regression model containing a distance variable. *Geographical Systems*, **3**, 77–99.
Okunuki, K. (2001). Urban analysis with GIS. *GeoJournal*, **52**, 181–188.
Openshaw, S. (1977a). Geographical solution to scale and aggregation problems in region-building, partitioning and spatial modeling. *Transactions of the Institute of British Geographers*, **2**, 459–472.
Openshaw, S. (1977b). Optimal zoning systems for spatial interaction models. *Environment and Planning A*, **9**, 169–184.
Openshaw, S. (1978a). An optimal zoning approach to the study of spatially aggregated data. In: I. Masser and P.J.B. Brown (eds.), *Spatial Representation and Spatial Interaction*, Martinus Nijhoff, Leiden, pp. 93–113.
Openshaw, S. (1978b). Empirical-study of some zone-design criteria. *Environment and Planning A*, **10**, 781–794.
Openshaw, S. (1984). *The Modifiable Areal Unit Problem*. Geo Books, Norwich.

References

Openshaw, S. (1991). A spatial analysis research agenda. In: I. Masser and M. Blakemore (eds.), *Handling Geographical Information: Methodology and Potential Applications*, John Wiley & Sons, New York, pp. 18–37.

Openshaw, S. (2000). Geocomputation. In: S. Openshaw and R.J. Abrahart (eds.), *Geocomputation*, Taylor and Francis, London.

Openshaw, S. and Alvandies, S. (1999). Applying geocomputation to the analysis of spatial distributions. In: P. Longley, M. Goodchild, D. Maguire and D. Rhind (eds.), *Geograhpic Information Systems: Principles and Technical Issues*, Vol.1, 2nd ed., John Wiley and Sons, New York.

Openshaw, S. and Baxter, R.S. (1977). Algorithm 3 – procedure to generate pseudo-random aggregations of n-zones into m-zones, where m is less than n. *Environment and Planning A*, **9**, 1423–1428.

Openshaw, S. and Rao, L. (1995). Algorithms for reengineering 1991 census geography. *Environment and Planning A*, **27**, 425–446.

Openshaw, S. and Taylor, P.J. (1979). A million or so correlation coefficients: three experiments on the modifiable areal unit problem. In: N. Wrigley (ed.), *Statistical Applications in the Spatial Sciences*, Pion, London, pp. 127–144.

Ord, J.K. and Getis, A. (1995). Local spatial autocorrelation statistics: distributional issues and an application. *Geographical Analysis*, **27**, 286–306.

Ortúzar, J.D. and Willumsen, L.G. (2001). *Modelling Transport*, 3rd ed. Wiley, New York.

OTA (1995). *The Technological Reshaping of Metropolitan America*. DCUS Congress Office of Technology Assessment, Washington.

Ottensmann, J.R. (1977). Urban sprawl, land values and the density of development. *Land Economics*, **53**(4), 389–400.

Oxford Dictionary (2000). *Oxford Advanced Learner's Dictionary of Current English*, 6th ed. Oxford University Press, Oxford.

Pacione, M. (1990). *Urban Problems: An Applied Urban Analysis*. Routledge, London.

Páez, A. and Scott, D.M. (2004). Spatial statistics for urban analysis: a review of techniques with examples. *GeoJournal*, **61**(1), 53–67.

Páez, A. and Suzuki, J. (2001). Transportation impacts on land use change: an assessment considering neighborhood effects. *Journal of the Eastern Asia Society for Transportation Studies*, **4**, 47–59.

Páez, A., Uchida, T. and Miyamoto, K. (2001). Spatial association and heterogeneity issues in land price models. *Urban Studies*, **38**, 1493–1508.

Pang, M.Y.C. and Shi, W.Z. (2002). Development of a process-based model for dynamic interaction in spatio-temporal GIS. *Geoinformatica*, **6**, 323–344.

Paola, J.D. and Schowengerdt, R.A. (1995). A detailed comparison of backpropagation neural network and maximum-likelihood classifiers for urban land use classification. *IEEE Transactions on Geoscience and Remote Sensing*, **33**, 981–996.

Paolini, L., Grings, F., Sobrino, J., Jiménez Muñoz, J. and Karszenbaum, H. (2006). Radiometric correction effects in Landsat multi-date/multi-sensor change detection studies. *International Journal of Remote Sensing*, **27**, 685–704.

Parker, D.C. (2004). Integration of geographic information systems and agent-based models of land use: challenges and prospects. In: D. Maguire, J.M.F. Goodchild, M. Batty (eds.), *GIS, Spatial Analysis and Modeling*, ESRI Press, Redlands, CA.

Parker, D.C., Evans, T.P. and Meretsky, V. (2001). Measuring emergent properties of agent-based landuse/ landcover models using spatial metrics. *Proceedings of 7th Annual Conference of the International Society for Computational Economics*. URL: http://php.indiana.edu/~dawparke/parker.pdf.

Parker, D.C., Manson, S.M., Janssen, M.A., Hoffmann, M.J. and Deadman, P. (2003). Multi-agent systems for the simulation of land-use and land-cover change: a review. *Annals of the Association of American Geographers*, **93**(2), 314–337.

Pathan, S.K., Jothimani, P., Pendharkar, S.P. and Kumar, D.S. (1989). Urban land-use mapping and zoning of Mombay Metropolitan Region using Remote Sensing data. *Journal of Indian Society of Remote Sensing*, **17**(3), 11–22.

Pathan, S.K., Shukla, V.K., Patel, R.G. and Mehta, K.S. (1991). Urban land-use mapping—a case study of Ahmedabad city and its environs. *Journal of Indian Society of Remote Sensing*, **19**(2), 95–112.

Patterson, J. (1999). *Urban Growth Boundary Impacts on Sprawl and Redevelopment in Portland, Oregon*. Working Paper, University of Wisconsin-Whitewater, WI.

Peddle, D.R., Hall, F.G. and Ledrew, E.F. (1999). Spectral mixture analysis and geometric-optical reflectance modeling of boreal forest biophysical structure. *Remote Sensing of Environment*, **67**(3), 288–297.

Pedersen, D., Smith, V.E. and Adler, J. (1999, July 19). Sprawling, sprawling. *Newsweek*, 23–27.

Peiser, R. (1989). Density and urban sprawl. *Land Economics*, **65**(3), 193–204.

Peiser, R.B. (1984). Does it pay to plan suburban growth? *Journal of the American Planning Association*, **50**(4), 419–433.

Pendall, R. (1999). Do land-use controls cause sprawl? *Environment and Planning B*, **26**(4), 555–571.

Pendall, R., Martin, J. and Fulton, W. (2002). *Holding the Line: Urban Containment in the United States*. The Brookings Institution Center on Urban and Metropolitan Policy, Washington, DC.

Peplies, R.W. (1973). Regional analysis and remote sensing: a methodological approach. In: J.E. Estes and L.W. Senger (eds.), *Remote Sensing: Techniques for Environmental Analysis*, Hamilton Pub. Co., Columbus, OH, pp. 277–291.

Pesaresi, M. and Bianchin, A. (2001). Recognizing settlement structure using mathematical morphology and image texture. In: J.P. Donnay, M.J. Barnsley and P.A. Longley (eds.), *Remote Sensing and Urban Analysis*, Taylor & Francis, London, pp. 56–67.

Phinn, S., Stanford, M., Scarth, P., Murray, A.T. and Shyy, P.T. (2002). Monitoring the composition of urban environments based on the vegetation-impervious surface-soil (VIS) model by subpixel analysis techniques. *International Journal of Remote Sensing*, **23**, 4131–4153.

Phipps, M. and Langlois, A. (1997). Spatial dynamics, cellular automata, and parallel processing computers. *Environment and Planning B*, **24**, 193–204.

Pijankowski, B.C., Long, D.T., Gage, S.H. and Cooper, W.E. (1997). A land transformation model: conceptual elements spatial object class hierarchy, GIS command syntax and an application for Michigan's Saginaw Bay watershed. *Proceedings of The Land Use Modeling Workshop*. June 5–6. URL: http://www.ncgia.ucsb.edu/conf/landuse97/.

Pijanowski, B.C., Brown, D.G., Shellitoc, B.A. and Manikd, G.A. (2002). Using neural networks and GIS to forecast land use changes: a land transformation model. *Computers, Environment and Urban Systems*, **26**(6), 553–575.

Pijanowski, B.C., Pithadia, S., Shellito, B.A. and Alexandridis, K. (2005). Calibrating a neural network-based urban change model for two metropolitan areas of the Upper Midwest of the United States. *International Journal of Geographical Information Science*, **19**, 197–215.

Pilon, P.G., Howarth, P.J. and Bullock, R.A. (1988). An enhanced classification approach to change detection in semi-arid environments. *Photogrammetric Engineering and Remote Sensing*, **54**(12), 1709–1716.

Pohl, C. and Van Genderen, J.L. (1998). Multisensor image fusion in remote sensing: Concepts, methods and applications. *International Journal of Remote Sensing*, **19**(5), 823–854.

Pollard, T. and Stanley, F. (2007). *Connections AND Choices: Affordable Housing and Smarter Growth in the Greater Richmond Area*. Southern Environmental Law Center, Charlottesville, VA. URL: http://www.southernenvironment.org/uploads/publications/richmond_housing_and_growth_Sept07.pdf.

Pooyandeh, M., Mesgari, S., Alimohammadi, A. and Shad, R. (2007). A comparison between complexity and temporal GIS models for spatio-temporal urban applications. In: O. Gervasi and M. Gavrilova (eds.), *ICCSA 2007, LNCS 4706, Part II*, Springer, pp. 308–321.

References

Popenoe, D. (1979). Urban sprawl: some neglected sociology. *Sociology and Social Research*, **31**(2), 181–188.

Portugali, J. (2000). *Self-organization and the City*. Springer, New York.

Portugali, J., Benenson, I. and Omer, I. (1997). Spatial cognitive dissonance and sociospatial emergence in a self-organizing city. *Environment and Planning B*, **24**, pp. 263–285.

Prenzel, B. (2004). Remote sensing-based quantification of land-cover and land-use change for planning. *Progress in Planning*, **61**, 281–299.

Priest, D. et al. (1977). *Large-Scale Development: Benefits, Constraints, and State and Local Policy Incentives*. Urban Land Institute, Washington, DC, pp. 37–45.

Puliafito, J.L. (2007). A transport model for the evolution of urban systems. *Applied Mathematical Modelling*, **31**, 2391–2411.

Putman, S.H. and Chung, S.H. (1989). Effects of spatial system-design on spatial interaction models.1. the spatial system definition problem. *Environment and Planning A*, **21**, 27–46.

Putnam, R.D. (2000). *Bowling Alone*. Simon and Schuster, New York.

Quandt, R. (1958). The estimation of parameters of a linear regression system obeying two separate regimes. *Journal of the American Statistical Association*, **53**, 873–880.

Quarmby, N.A. and Cushnie, J.L. (1989). Monitoring urban land cover changes at the urban fringe from SPOT HRV imagery in south-east England. *International Journal of Remote Sensing*, **10**, 953–963.

Quattrochi, D.A. (1983). Analysis of Landsat-4 thematic mapper data for classification of the Mobile, Alabama metropolitan area. *Proceedings of 17th International Symposium on Remote Sensing of Environment*. ERIM (Ed.: M.I. Ann Arbor), pp. 1393–1399.

Quenzel, H. (1983). Principles of remote sensing techniques. In: P. Camagni and S. Sandroni (eds.), *Optical Remote Sensing of Air Pollution*, Elsevier Science, Amsterdam, pp. 27–43.

Radeloff, V.C., Holcomb, S.S., McKeefry, J.F., Hammer, R.B. and Stewart, S.I. (2005). The wildland-urban interface in the United States. *Ecological Applications*, **15**, 799–805.

Rahman, G., Alam, D. and Islam, S. (2008). City growth with urban sprawl and problems of management. *Proceedings of 44th ISOCARP Congress Dalian, China*, September 19–23, International Society of City and Regional Planners and Urban Planning Society of China.

Ramos, R.A.R. and Silva, A.N.R. (2003). *A data-driven approach for the definition of metropolitan regions*. CUPUM (CD-ROM), Sendai, Japão: Proceedings of the VIII International Conference on Computers in Urban Planning and Urban Management.

Rathbone, D.B. and Huckabee, J.C. (1999). *Controlling Road Rage: A Literature Review and Pilot Study*. AAA Foundation for Traffic Safety; Washington, DC.

Ray, T.W. and Murray, B.C. 1996. Non-linear spectral mixing in desert vegetation. *Remote Sensing of Environment*, **55**(1), 59–64.

RERC (Real Estate Research Corporation) (1974). *The costs of sprawl: Environmental and economic costs of alternative residential patterns at the Urban Fringe*. US Government Printing Office, Washington, DC.

Ridd, M.K. and Liu, J.J. (1998). A comparison of four algorithms for change detection in an urban environment. *Remote Sensing of Environment*, **63**, 95–100.

Rigol, J.P., Jarvis, C.H. and Stuart, N. (2001). Artificial neural networks as a tool for spatial interpolation. *International Journal of Geographical Information Science*, **15**(4), 322–343.

Riitters, K.H., O'Neill, R.V., Hunsaker, C.T., Wickham, J.D., Yankee, D.H., Timmins, S.P., Jones, K.B. and Jackson, B.L. (1995). A factor analysis of landscape pattern and structure metrics. *Landscape Ecology*, **10**(1), 23–39.

Riordan, C.J. (1980). *Non-urban to Urban Land Cover Change Detection Using Landsat Data, Summary Report of the Colorado Agricultural Research Experiment Station*, Fort Collins, CO.

Robbins, P. and Birkenholtz, T. (2003). Turfgrass revolution: measuring the expansion of the American Lawn. *Land Use Policy*, **20**, 181–194.

Robinson, J.W. (1979). *A Critical Review of the Change Detection and Urban Classification Literature, Technical Memorandum CSC/TM-79/6235*. Computer Sciences Corporation, Silver Springs, MD.

Robinson, T.P. (2000). Spatial statistics and geographical information systems in epidemiology and public health. *Advances in Parasitology*, **47**, 81–128.

Roca, J., Burnsa, M.C. and Carreras, J.M. (2004). Monitoring urban sprawl around Barcelona's metropolitan area with the aid of satellite imagery. *Proceedings of Geo-Imagery Bridging Continents*, XXth ISPRS Congress, July 12–23, Istanbul, Turkey.

Rodrigue, J.P. (1997). Parallel modeling and neural networks: an overview for transportation/land use systems. *Transportation Research Part C*, **5**(5), 259–271.

Rogan, J., Franklin, J. and Roberts, D.A. (2002). A comparison of methods for monitoring multi-temporal vegetation change using Thematic Mapper imagery. *Remote Sensing of Environment*, **80**, 143–156.

Rosin, P.L. (2002). Thresholding for change detection. *Computer Vision and Image Understanding*, **86**, 79–95.

Rosin, P.L. and Ioannidis, E. (2003). Evaluation of global image thresholding for change detection. *Pattern Recognition Letters*, **24**, 2345–2356.

Rouse, J.W., Haas, R.H., Schell, J.A., Deering, D.W. and Harlan, J.C. (1974). Monitoring the Vernal Advancement and Retrogradation (Greenwave Effect) of Natural Vegetation. Remote Sensing Center Report RSC 1978-4, Texas A&M University, College Station TX.

Sabins, F.F. (1996). *Remote Sensing: Principles and Interpretation*. W. H. Freeman, New York.

Sadowski, F.G., Sturdevant, J.A. and Rowntree, R.A. (1987). Testing the consistency for mapping urban vegetation with high-altitude aerial photographs and Landsat MSS data. *Remote Sensing of the Environment*, **21**, 129–141.

Sakashita, N. (1995). An economic theory of urban growth control. *Regional Science and Urban Economics*, **25**, 427–434.

Salem, B.B., El-Cibahy, A. and El-Raey, M. (1995). Detection of land cover classes in agro-ecosystems of northern Egypt by remote sensing. *International Journal of Remote Sensing*, **16**(14), 2581–2594.

Sanders, L., Pumain, D., Mathian, H., Guérin-Pace, F. and Bura, S. (1997). SIMPOP: a multi-agent system for the study of urbanism. *Environment and Planning B*, **24**, 287–305.

Sasaki, K. (1998). Optimal urban growth controls. *Regional Science and Urban Economics*, **28**(4), 475–496.

Savitch, H.V. (2003). How suburban sprawl shapes human well-being. *Journal of Urban Health: Bulletin of the New York Academy of Medicine*, **80**(4), 590–607.

Schelhorn, T., O'Sullivan, D., Haklay, M. and Thurstain-Goodwin, M. (1999). *Streets: An Agentbased Pedestrian Model*. Working Paper Series, Paper 9, Centre for Advanced Spatial Analysis, University College London.

Schiffman, I. (1999). *Alternative Techniques for Managing Growth*, 2nd Edition. Institute of Governmental Studies Press, University of California, Berkeley.

Schneider, A., Seto, K.C. and Webster, D.R. (2005). Urban growth in Chengdu, Western China: applications of remote sensing to assess planning and policy outcomes. *Environment and Planning B*, **32**, 323–345.

Schott, J. (1997). *Remote Sensing: the Image Chain Approach*. Oxford University Press, New York.

Schott, J.R., Salvaggio, C. and Volchock, W.J. (1988). Radiometric scene normalization using pseudoinvariant features. *Remote Sensing of Environment*, **26**, 1–16.

Schowengerdt, R.A. (1983). *Techniques of Image Processing and Classification in Remote Sensing*. Academic Press, New York.

Schowengerdt, R.A. (1997). *Remote Sensing: Models and Methods for Image Processing*. Academic Press, New York.

Schweitzer, F. (2002). Modelling migration and economic agglomeration with active brownian particles. In: F. Schweitzer (ed.), *Modeling Complexity in Economic and Social Systems*, World Scientific, Singapore.

Schweitzer, F. (2003). *Brownian Agents and Active Particles: Collective Dynamics in the Natural and Social Sciences*. Springer, Berlin.

Schweitzer, F. and Holyst, J. (2000). Modelling collective opinion formation by means of active brownian particles. *European Physical Journal B*, **15**(4), 723–732.

References

Schweitzer, F. and Tilch, B. (2002). Self-assembling of networks in an agent-based model. *Physical Review E*, **66**(2), 1–9.

SCN (Sustainable Communities Network) (2009). *About Smart Growth*. URL: http://www.smartgrowth.org/about/default.asp.

Segal, D. and Srinivasan, P. (1985). The impact of suburban growth restrictions on U.S. housing price inflation. *Urban Geography*, **6**, 14–26.

Semboloni, F. (1997). An urban and regional model based on cellular automata. *Environment and Planning B*, **24**(4), 589–612.

Semboloni, F., Assfalg, J., Armeni, J., Gianassi, R. and Marsoni, F. (2004). CityDev, an interactive multi-agents urban model on the web. *Computers, Environment and Urban Systems*, **28**, 45–64.

Serra, P., Pons, X. and Sauri, D. (2003). Post-classification change detection with data from different sensors: some accuracy considerations. *International Journal of Remote Sensing*, **24**, 3311–3340.

Seto, K.C. and Fragkias, M. (2005). Quantifying spatiotemporal patterns of urban land-use change in four cities of China with timer series landscape metrics. *Landscape Ecology*, **20**, 871–888.

Shen, G. (2002). Fractal dimension and fractal growth of urbanized areas. *International Journal of Geographical Information Science*, **16**(5), 419–437.

Shmueli, D. (1998). Applications of neural networks in transportation planning. *Progress in Planning*, **50**, 141–204.

Sierra Club (1998). *Sprawl: the Dark Side of the American Dream*. Research Report. URL: http://www.sierraclub.org/sprawl/ report98/report.asp.

Sierra Club (2001). *Stop Sprawl: New Research on Population, Suburban Sprawl, and Smart Growth*. URL: www.sierraclub.org/sprawl/.

Silva, E.A. and Clarke, K.C. (2002). Calibration of the SLEUTH urban growth model for Lisbon and Porto, Portugal. *Computers, Environment and Urban Systems*, **26**, 525–552.

Singh, A. (1989). Digital change detection techniques using remotely sensed data. *International Journal of Remote Sensing*, **10**(6), 989–1003.

Singh, A. and Harrison, A. (1985). Standardized principal components. *International Journal of Remote Sensing*, **6**, 883–896.

Smart Growth Network (2003). *Getting to Smart Growth: 100 Policies for Implementation*. International City/County Management Organization (ICMA).

Smith, D.M. (1975). *Patterns in Human Geography: David & Charles*. Newton Abbot, England, p. 373.

Smits, P.C. and Annoni, A. (2000). Toward specification-driven change detection. *IEEE Transactions on Geoscience and Remote Sensing*, **38**, 1484–1488.

Smolensky, P. (1998). On the proper treatment of connectionism. *Behavioral and Brain Sciences*, **11**, 1–74.

Soares-Filho, B.S., Coutinho-Cerqueira, G. and Lopes-Pennachin, C. (2002). DINAMICA—A stochastic cellular automata model designed to simulate the landscape dynamics in an Amazonian colonization frontier. *Ecological Modeling*, **154**(3), 217–235.

Song, C., Woodcock, C.E., Seto, K.C., Pax-Lenney, M. and Macomber, S.A. (2001). Classification and change detection using Landsat TM data: when and how to correct for atmospheric effects? *Remote Sensing of Environment*, **75**, 230–244.

Spiekermann, K. and Wegener, M. (2000). Freedom from the tyranny of zones: towards new GIS-based spatial models. In: A.S. Fotheringham and M. Wegener (eds.), *Spatial Models and GIS: New Potential and New Models*, Taylor and Francis, London, pp. 45–61.

Squires, G.D. (2002). *Urban Sprawl Causes, Consequences and Policy Responses*. Urban Institute Press, Washington, DC.

Staley, S.R. and Gilroy, L.C. (2004). *Smart growth and housing affordability: The academic evidence*. The Brookings Institution Centre on Urban and Metropolitan Policy, Washington, DC.

Staley, S.R., Edgens, J.G. and Mildner, G.C.S. (1999). *A Line in the Land: Urban-Growth Boundaries, Smart Growth, and Housing Affordability*, Policy Study No. 263. Los Angeles, CA. URL: http://www.reason.org/ps263.html.

Stanilov, K. (2003). Accessibility and land use: the case of suburban Seattle, 1960–1990. *Regional Studies*, **37**, 783–794.

Steel, D.G. and Holt, D. (1996). Rules for random aggregation. *Environment and Planning A*, **28**, 957–978.

Steenberghen, T., Dufays, T., Thomas, I. and Flahaut, B. (2004). Intra-urban location and clustering of road accidents using GIS: a Belgian example. *International Journal of Geographical Information Science*, **18**, 169–181.

Stoel Jr., T.B. (1999). Reining in urban sprawl. *Environment*, **41**(4), 6–33.

Stone Jr., B. (2008). Urban sprawl and air quality in large US cities. *Journal of Environmental Management*, **86**, 688–698.

Stone, P.A. (1973). *The Structure, Size, and Costs of Urban Settlements*. Cambridge University Press, London.

Stow, D.A., Tinney, L.R. and Estes, J.E. (1980). Deriving land use/land cover change statistics from Landsat: a study of prime agricultural land. *Proceedings of the 14th International Symposium on Remote Sensing of the Environment*, Environmental Research Institute of Michigan, Ann Arbor, MI, pp. 1227–1237.

Sturm, R. and Cohen, D.A. (2004). Suburban sprawl and physical and mental health. *Public Health*, **118**, 488–496.

Sudhira, H.S. and Ramachandra, T.V. (2007). Characterising urban sprawl from remote sensing data and using landscape metrics. *Proceedings of 10th International Conference on Computers in Urban Planning and Urban Management*, Iguassu Falls, PR Brazil, July 11–13. URL: http://eprints.iisc.ernet.in/11834/.

Sudhira, H.S., Ramachandra, T.V. and Jagdish, K.S. (2004). Urban sprawl: metrics, dynamics and modelling using GIS. *International Journal of Applied Earth Observation and Geoinformation*, **5**, 29–39.

Sui, D.Z. (1998). GIS-based urban modeling: practices, problems, and prospects. *International Journal of Geograhical Information Science*, **12**(7), 651–671.

Sunar, F. (1998). Analysis of changes in a multidate set: a case study in Ikitelli Area, Istanbul, Turkey. *International Journal of Remote Sensing*, **19**, 225–235.

Sutton, P.C. (2003). A scale-adjusted measure of "urban sprawl" using nighttime satellite imagery. *Remote Sensing of Environment*, **86**, 353–369.

Swain, P.H. (1978). Bayesian classification in a time-varying environment. *IEEE Transactions on Systems, Man and Cybernetics*, **8**, 879–883.

Syphard, A.D., Clarke, K.C. and Franklin, J. (2005). Using a cellular automaton model to forecast the effects of urban growth on habitat pattern in southern California. *Ecological Complexity*, **2**, 185–203.

Tagashira, N. and Okabe, A. (2002). The modifiable areal unit problem in a regression model whose independent variable is a distance from a predetermined point. *Geographical Analysis*, **34**, 1–20.

Taket, N.D., Howarth, S.M. and Burge, R.E. (1991). A model for the imaging of urban areas by synthetic aperture radar. *IEEE Transactions on Geoscience and Remote Sensing*, **29**, 432–443.

Taylor, P.J. (1977). *Quantitative Methods in Geography: An Introduction to Spatial Analysis*. Houghton Mifflin Company, Boston, Massachusetts.

The Brookings Institution (2002). *The Growth in the Heartland: Challenges and Opportunities for Missouri*. The Brookings Institution Center on Urban and Metropolitan Policy, Washington, DC.

Theil, H. (1967). *Economics and Information Theory*. North-Holland, Amsterdam, p. 488.

Theobald, D.M. (2005). Landscape patterns of exurban growth in the USA from 1980 to 2020. *Ecology and Society*, **10**(1), article 32. URL: http://www.ecologyandsociety.org/vol10/iss1/art32/.

Thomas, R.W. (1981). *Information Statistics in Geography, Geo Abstracts*. University of East Anglia, Norwich, p. 42.

References

Tobler, W. (1970). A computer movie simulating urban growth in the Detroit region. *Economic Geography*, **46**, 234–240.

Tobler, W.R. (1989). Frame independent spatial analysis. In: M. Goodchild and S. Gopal (eds.), *Accuracy of Spatial Databases*, Taylor & Francis, London, pp. 115–122.

Todd, W.J. (1977). Urban and regional land use change detected by using Landsat data. *Journal of Research by the U.S. Geological Survey*, **5**, 529–34.

Toll, D.L. (1980). Urban area update procedures using Landsat data. *Proceedings of the American Society of Photogrammetry*, 5410 Grosvenor Lane, Bethesda, MD 20814–2160, RS-1-E-1-RS-1-E-17.

Toll, D.L., Royal, J.A. and Davis, J.B. (1984). An evaluation of simulated Thematic Mapper data and Landsat MSS data for discriminating suburban and regional land use and land cover. *Photogrammetric Engineering and Remote Sensing*, **50**, 1713–1724.

Torrens, P. (2003). Automata-based models of urban systems. In: M. Batty and P. Longley (eds.), *Advanced Spatial Analysis*, ESRI Press, Redlands.

Torrens, P.M. (2000). *How Cellular Models of Urban Systems Work (1. Theory)*. Working Paper Series 28, Centre for Advanced Spatial Analysis, London, UK. URL: http://eprints.ucl.ac.uk/1371/.

Torrens, P.M. (2001). *Can Geocomputation Save Urban Simulation? Throw Some Agents into the Mixture, Simmer, and Wait*. Paper 32, Centre for Advanced Spatial Analysis, University College London.

Torrens, P.M. (2006). Geosimulation and its application to urban growth modeling. In: J. Portugali (ed.), *Complex Artificial Environments*, Springer-Verlag, London, pp. 119–134.

Torrens, P.M. (2008). A toolkit for measuring sprawl. *Applied Spatial Analysis*, **1**, 5–36.

Torrens, P.M. and Benenson, I. (2005). Geographic Automata Systems. *International Journal of Geographic Information Systems*, **19**(4), 385–412.

Torrens, P.M. and O'Sullivan, D. (2001). Cellular automata and urban simulation: where do we go from here? *Environment and Planning B*, **28**, 163–168.

Townshend, J.R.G. and Justice, C.O. (1995). Spatial variability of images and the monitoring of changes in the normalized difference vegetation index. *International Journal of Remote Sensing*, **16**, 2187–2195.

Tranmer, M. and Steel, D.G. (1998). Using census data to investigate the causes of the ecological fallacy. *Environment and Planning A*, **30**, 817–831.

Tregoning, H., Agyeman, J. and Shenot, C. (2002). Sprawl, smart growth and sustainability. *Local Environment*, **7**(4), 341–347.

Tsai, Y. (2005). Quantifying urban form: compactness versus 'sprawl'. *Urban Studies*, **42**(1), 141–161.

Tse, R.Y.C. (2002). Estimating neighbourhood effects in house prices: towards a new hedonic model approach. *Urban Studies*, **39**, 1165–1180.

Turner, M.G. (1989). Landscape ecology: the effects of pattern on process. *Annual Review of Ecological Systems*, **20**, 171–197.

Turner, M.G., Gardner, R.H. and O'Neill, R.V. (2001). *Landscape Ecology in Theory and Practice: Pattern and Process*. Springer, New York, p. 401.

U.S. Census Bureau (2000). *Urban and Rural Classification*. URL: http://www.census.gov/geo/www/ua/ua 2 k.html.

UNFPA (United Nations Population Fund) (2007). *Peering into the Dawn of an Urban Millennium, State of World Population 2007: Unleashing the Potential of Urban Growth*. URL: www.unfpa.org/swp/2007/english/introduction.html.

United Nations (1994). *World Urbanization Prospects*. Population Division, United Nations, New York.

United Nations (2002). *World Urbanisation Prospects, 2002 Revision*. Population Division, United Nations, New York.

United Nations (2005a). *World Urbanization Prospects: The 2005 Revision*. Pop. Division, Department of Economic and Social Affairs, UN. URL: http://www.un.org/esa/population/publications/WUP2005/2005wup.htm.

United Nations (2005b). *World Summit Outcome Document*. World Health Organization, United Nations.

United Nations (2009). *The Millennium Development Goals Report 2009*. United Nations. URL: http://www.un.org/millenniumgoals.

USA Today (2001). *A Comprehensive Look at Sprawl in America*. URL: http://www.usatoday.com/news/sprawl/ main.htm.

USEPA (United States Environmental Protection Agency) (2001). *Our Built and Natural Environments*. A Technical Review of the Interactions between Land Use, Transportation, and Environmental Quality. US Environmental Protection Agency. URL: http://www.smartgrowth.org/pdf/built.pdf.

USEPA (United States Environmental Protection Agency) (2009). *About Smart Growth*. US Environmental Protection Agency. URL: http://www.epa.gov/smartgrowth/about_sg.htm.

USGAO (United States General Accounting Office) (1999). *Community Development: Extent of Federal Influence on "Urban Sprawl" Is Unclear*. Letter Report, Washington, DC.

Usher, J.M. (2000). *Remote Sensing Applications in Transportation Modeling*. Remote Sensing Technology Centers Final Report. URL: http://www.rstc.msstate.edu/publications/proposal1999-2001.html.

Uy, P.D. and Nakagoshi, N. (2008). Application of land suitability analysis and landscape ecology to urban greenspace planning in Hanoi, Vietnam. *Urban Forestry & Urban Greening*, 7, 25–40.

Verbyla, D.L. and Richardson, C.A. (1996). Remote sensing clearcut areas within a forested watershed: comparing SPOT HRV panchromatic, SPOT HRV multispectral, and Landsat Thematic Mapper data. *Journal of Soil and Water Conservation*, 51(5), 423–427.

Vogelmann, J.E., Sohl, T. and Howard, S.M. (1998). Regional characterization of land cover using multiple sources of data. *Photogrammetric Engineering and Remote Sensing*, 64, 45–57.

Waddell, P. (1998). UrbSim—the oregon prototype metropolitan land use model. *Proceedings of the ASCE Conference Transportation, Land Use, and Air Quality: Making the Connection Portland, Oregon*, May 1998. URL: http://www.urbansim.org/Papers/ASCE%20Model.pdf.

Waddell, P. (2002). UrbanSim: modelling urban development for land use, transportation and environmental planning. *Journal of the American Planning Association*, 68, 297–314.

Waddell, P. and Evans, D. (2002). *UrbanSim: Modelling Urban Land Development for Land Use*. Transportation and Environmental Planning, University of Washington, Seattle.

Wang, D.G. and Cheng, T. (2001). A spatio-temporal data model for activity-based transport demand modelling. *International Journal of Geographical Information Science*, 15, 561–585.

Wang, W., Zhu, L., Wang, R. and Shi, Y. (2003). Analysis on the spatial distribution variation characteristic of urban heat environmental quality and its mechanism—a case study of Hangzhou city. *Chinese Geographical Science*, 13(1), 39–47.

Wang, Y., Jamshidi, M., Neville, P., Bales, C. and Morain, S. (2007). Hierarchical fuzzy classification of remote sensing data. *Studies in Fuzziness and Soft Computing*, 217, 333–350.

Ward, D., Phinn, S.R. and Murray, A.T. (2000). Monitoring growth in rapidly urbanizing areas using remotely sensed data. *Professional Geographer*, 52, 371–386.

Wasserman, M. (2000). Confronting urban sprawl. *Regional Review of the Federal Reserve Bank of Boston*, pp. 9–16.

Wassmer, R.W. (2002). Fiscalisation of land use, urban growth boundaries and non-central retail sprawl in the western United States. *Urban Studies*, 39(8), 1307–1327.

Wassmer, R.W. and Baass, M.C. (2006). Does a more centralized urban form raise housing prices? *Journal of Policy Analysis and Management*, 25, 439–462.WCED (World Commission on Environment and Development) 1987. *Our Common Future*. Oxford University Press, Oxford.

Weber, C. and Puissant, A. (2003). Urbanization pressure and modeling of urban growth: example of the Tunis Metropolitan area. *Remote Sensing of Environment*, 86, 341–352.

Webster, C.J. (1995). Urban morphological fingerprints. *Environment and Planning B*, 22, 279–297.

Webster, C.J. (1996). Urban morphology fingerprints. *Environment and Planning B*, 23, 279–297.

References

Wegener, M. (1994). Operational urban models: state of the art. *Journal of the American Planning Association*, **60**(1), 17–30.

Weiler, S. and Theobald, D. (2003). Pioneers of rural sprawl in the Rocky Mountain West. *Review of Regional Studies*, **33**, 264–283.

Weismiller, R.A., Kristof, S.J., Scholz, D.K., Anuta, P.E. and Momen, S.A. (1977). Change detection in coastal zone environments. *Photogrammetric Engineering and Remote Sensing*, **43**, 1533–1539.

Weitz, J. (1999). *Sprawl Busting: State Programs to Guide Growth*. APA Planners Press, Chicago, IL.

Weitz, J. and Moore, T. (1998). Development inside urban growth boundaries: oregon's empirical evidence of contiguous urban form. *Journal of the American Planning Association*, **64**(4), 424–440.

Weng, Q. (2001). A remote sensing-GIS evaluation of urban expansion and its impact on surface temperature in the Zhujiang Delta, China. *International Journal of Remote Sensing*, **22**(10), 1999–2014.

Weng, Q., Liu, H. and Lu, D. (2007). Assessing the effects of land use and land cover patterns on thermal conditions using landscape metrics in city of Indianapolis, United States. *Urban Ecosystem*, **10**, 203–219.

White, R. and Engelen, G. (1993). Cellular automata and fractal urban form: a cellular modeling approach to the evolution of urban land-use patterns. *Environment and Planning A*, **25**, 1175–1199.

White, R. and Engelen, G. (1997). Cellular automata as the basis of integrated dynamic regional modelling. *Environment and Planning B: Planning and Design*, **24**(2), 235–246.

White, R. and Engelen, G. (2000). High resolution integrated modeling of the spatial dynamics of urban and regional systems. *Computers, Environment and Urban Systems*, **24**, 383–440.

White, R.W., Engelen, G. and Uljee, I. (1998). *Vulnerability Assessment of Low-Lying Coastal Areas and Small Islands to Climate Change and Sea Level Rise – Phase 2: Case Study St. Lucia*. Report to the United Nations Environment Programme, Caribbean Regional Co-ordinating Unit, RIKS Publication, Kingston, Jamaica.

Willson, R.W. (1995). Suburban parking requirements: a tacit policy for automobile use and sprawl. *Journal of the American Planning Association*, **61**, 29–42.

Wilson, A.G. (2000). *Complex Spatial Systems: The Modeling Foundations of Urban and Regional Analysis*. Pearson Education, London and Harlow.

Wilson, E.H., Hurd, J.D., Civco, D.L., Prisloe, S. and Arnold, C. (2003). Development of a geospatial model to quantify, describe and map urban growth. *Remote Sensing of Environment*, **86**(3), 275–285.

Wisner, B., Blaikie, P., Cannon, T. and Davies, I. (2004). *At Risk, Natural Hazards, People's Vulnerability and Disasters*. Routlegde, London, p. 471.

Wong, D.W.S. (1996). Aggregation effects in geo-referenced data. In: S.L. Arlinghaus, D.A. Griffith, W.D. Drake and J.D. Nystuen (eds.), *Practical Handbook of Spatial Statistics*, CRC Press, Boca Raton, pp. 83–106.

Wong, D.W.S., Lasus, H. and Falk, R.F. (1999). Exploring the variability of segregation index D with scale and zonal systems: an analysis of thirty US cities. *Environment and Planning A*, **31**, 507–522.

Woodcock, C.E. and Strahler, A.H. (1987). The factor scale in remote sensing. *Remote Sensing of Environment*, **21**, 311–332.

WSSD (2002). *World Summit on Sustainable Development, Johannesburg*, August 26–September 4. URL: http://www.worldsummit2002.org.

Wu, F. (1998). An experiment on the generic polycentricity of urban growth in a cellular automatic city. *Environment and Planning B*, **25**, 731–752.

Wu, F. (1999). A simulation approach to urban changes: experiments and observations on fluctuations in cellular automata. In: P. Rizzi (ed.), *Proceedings of Sixth International Conference on Computers in Urban Planning and Urban Management*, Venice, Italy.

Wu, F. and Martin, D. (2002). Urban expansion simulation of Southeast England using population surface modeling and cellular automata. *Environment and Planning A*, **34**(10), 1855–1876.

Wu, F. and Webster, C.T. (1998). Simulation of land development through the integration of cellular automata and multi-criteria evaluation. *Environment and Planning B*, **25**, 103–126.

Wu, J., Jelinski, E.J., Luck, M. and Tueller, P.T. (2000). Multiscale analysis of landscape heterogeneity: scale variance and pattern metrics. *Geographic Information Sciences*, **6**(1), 6–16.

Xian, G. and Crane, M. (2005). Assessments of urban growth in the Tampa Bay watershed using remote sensing data. *Remote Sensing of Environment*, **97**(2), 203–215.

Xiao, J., Shen, Y., Ge, J., Tateishi, R., Tang, C., Liang, Y. and Huang, Z. (2006). Evaluating urban expansion and land use change in Shijiazhuang, China, by using GIS and remote sensing. *Landscape and Urban Planning*, **75**, 69–80.

Xie, Y. and Batty, M. (1997). Automata-based exploration of emergent urban form. *Geographical System*, **4**, 83–102.

Yang, Q., Li, X. and Shi, X. (2008). Cellular automata for simulating land use changes based on support vector machines. *Computers and Geosciences*, doi:10.1016/j.cageo.2007.08.003.

Yang, X. (2002). Satellite monitoring of urban spatial growth in the Atlanta metropolitan area. *Photogrammetric Engineering and Remote Sensing*, **68**, 725–734.

Yang, X. and Lo, C.P. (2002). Using a time series of satellite imagery to detect land use and cover changes in the Atlanta, Georgia. *International Journal of Remote Sensing*, **23**(9), 1775–1798.

Yang, X. and Lo, C.P. (2003). Modelling urban growth and landscape changes in the Atlanta metropolitan area. *International Journal of Geographical Information Science*, **17**, 463–488.

Yanos, P.T. (2007). Beyond "Landscapes of despair": the need for new research on the urban environment, sprawl, and the community integration of persons with severe mental illness. *Health & Place*, **13**, 672–676.

Yeh, A.G.O. and Li, X. (1997). An integrated remote sensing and GIS approach in the monitoring and evaluation of rapid urban growth for sustainable development in the Pearl River Delta, China. *International Planning Studies*, **2**, 193–210.

Yeh, A.G.O. and Li, X. (2001a). Measurement and monitoring of urban sprawl in a rapidly growing region using entropy. *Photogrammetric Engineering and Remote Sensing*, **67**(1), 83–90.

Yeh, A.G.O. and Li, X. (2001b). A constrained CA model for the simulation and planning of sustainable urban forms by using GIS. *Environment and Planning B*, **28**, 733–753.

Yeh, A.G.O. and Li, X. (2002). A cellular automata model to simulate development density for urban planning. *Environment and Planning B*, **29**, 431–450.

Yeh, A.G.O. and Li, X. (2003). Uncertainties in urban simulation using cellular automata and GIS. *Proceedings of the 7th International Conference on GeoComputation*, University of Southampton, UK, September 8–10. URL: http://www.geocomputation.org/2003/index.html.

Yeh, A.G.O. and Li, X. (2004). Integration of Neural Networks and Cellular Automata for Urban Planning. *Geo-Spatial Information Science*, **7**(1), 6–13.

You, J., Nedoviæ-Budiæ, Z. and Kim, T.J. (1997a). A GIS-based traffic analysis zone design: technique. *Transportation Planning and Technology*, **21**, 45–68.

You, J.S., Nedoviæ-Budiæ Z. and Kim T.J. (1997b). A GIS-based traffic analysis zone design: implementation and evaluation. *Transportation Planning and Technology*, **21**, 69–91.

Yu, S. and Berthod, M. and Giraudon, G. (1999). Toward robust analysis of satellite images using map information—application to urban area detection. *IEEE Transactions on Geoscience and Remote Sensing*, **37**, 1925–1939.

Yu, X. and Ng, C. (2006). An integrated evaluation of landscape change using remote sensing and landscape metrics: a case study of Panyu, Guangzhou. *International Journal of Remote Sensing*, **27**, 1075–1092.

Yu, X. and Ng, C. (2007). Spatial and temporal dynamics of urban sprawl along two urban–rural transects: A case study of Guangzhou, China. *Landscape and Urban Planning*, **79**, 96–109.

Yuan, D., Elvidge, C.D. and Lunetta, R. S. (eds.) (1999). *Survey of Multispectral Methods for Land Cover Change Analysis*. Taylor and Francis, London.

References

Yuan, F. (2008). Land-cover change and environmental impact analysis in the Greater Mankato area of Minnesota using remote sensing and GIS modelling. *International Journal of Remote Sensing*, **29**(4), 1169–1184.

Yuan, F., Sawaya, K.E., Loeffelholz, B.C. and Bauer, M.E. (2005). Land cover classification and change analysis of the Twin Cities (Minnesota) Metropolitan Area by multi-temporal Landsat remote sensing. *Remote Sensing of Environment*, **98**(2&3), 317–328.

Yue, W., Xu, J., Wu, J. and Xu, L. (2006). Remote sensing of spatial patterns of urban renewal using linear spectral mixture analysis: a case of central urban area of Shanghai (1997—2000). *Chinese Science Bulletin*, **51**(8), 977–986.

Zarco-Tejada, P. and Miller, J. (1999). Land cover mapping at BOREAS using red edge spectral parameters from CASI imagery. *Journal of Geophysical Research*, **104**(D22), 27921–27933.

Zarco-Tejada, P.J. (2000). Hyperspectral remote sensing of closed forest canopies: estimation of chlorophyll fluorescence and pigment content. *Department of Physics*, York University, Toronto, p. 210.

Zarco-Tejada, P.J., Miller, J.R., Noland, T.L., Mohammed, G.H. and Sampson, P.H. (2001). Scaling-up and model inversion methods with narrowband optical indices for chlorophyll content estimation in closed forest canopies with hyperspectral data. *IEEE Transactions on Geoscience and Remote Sensing*, **39**(7), 1491–1507.

Zeilhofer, P. and Topanotti, V.P. (2008). GIS and ordination techniques for evaluation of environmental impacts in informal settlements: a case study from Cuiaba, Central Brazil. *Applied Geography*, **28**, 1–15.

Zha, Y., Gao, J. and Ni, S. (2003). Use of normalized difference built-up index in automatically mapping urban areas from TM imagery. *International Journal of Remote Sensing*, **24**, 583–594.

Zhang, B. (2003). Application of remote sensing technology to population estimation. Chinese Geographical Science, **13**(3), 267–271.

Zhang, B. (2004). *Study on Urban Growth Management in China*. Xinhua Press, Beijing.

Zhang, H. (1997). A simulation of the dynamics of soil erosion in the loess hills of Shanxi and Shaanxi provinces. *Chinese Science Bulletin*, **42**(7), 743–746.

Zhang, J. and Foody, G.M. (1998). A fuzzy classification of sub-urban land cover from remotely sensed imagery. *International Journal of Remote Sensing*, **19**, 2721–2738.

Zhang, P. and Atkinson, P.M. (2008). Modelling the effect of urbanization on the transmission of an infectious disease. *Mathematical Biosciences*, **211**, 166–185.

Zhang, X., Chen, J., Tan, M. and Sun, Y. (2007). Assessing the impact of urban sprawl on soil resources of Nanjing city using satellite images and digital soil databases. *Catena*, **69**, 16–30.

Zhang, Y. and Guindon, B. (2006). Using satellite remote sensing to survey transport-related urban sustainability: Part 1: Methodologies for indicator quantification. *International Journal of Applied Earth Observation and Geoinformation*, **8**(3), 149–164.

Zhang, Y., Yang, Z. and Li, W. (2006). Analyses of urban ecosystem based on information entropy. *Ecological Modelling*, **197**, 1–12.

Zhu, M., Xu, J., Jiang, N., Li, J. and Fan, Y. (2006). Impacts of road corridors on urban landscape pattern: a gradient analysis with changing grain size in Shanghai, China. *Landscape Ecology*, **21**, 723–734.

Zilans, A. and Abolina, K. (2009). A methodology for assessing urban sustainability: aalborg commitments baseline review for Riga, Latvia. *Environment Development and Sustainability*, **11**(1), 85–114.

Zipperer, W.C., Sisinni, S.M., Pouyat, R.V. and Foresman, T.W. (1997). Urban tree cover: an ecological perspective. *Urban Ecosystems*, **1**, 229–246.

Index

A
Affordable housing, 18, 23, 40, 42
Agent, 113
Agent-based model(ling), 90, 95, 109, 112–114
Agricultural
　land, 31
　zoning, 45
Air quality, 33
Annexation, 46
Artificial neural network, 66, 76, 95, 114
Autocorrelation, 92–93, 97, 131

B
Built-up, 4

C
Capital market, 18, 26
CAST, 117
Cell-based dynamics, 111
Cellular automata, 95, 109, 111, 117
Centrality, 91, 99
Change
　detection, 63, 65, 66
　image, 68–69, 91
　matrix, 80
　vector analysis, 66, 75
City, 6
Class metrics, 90
Classification
　accuracy, 120, 126–127
　image, 51, 53, 62, 77, 79–82, 125–127
　land, 4
　spectral-temporal, 66, 72
　urban model, 109
Clustered branch, 10–12
Clustering, 13, 80, 93, 95, 99, 131
Coalescence, 15
Compactness, 9, 13–14, 91, 101–102
Complex adaptive system, 95

Complexity
　science, 110–116
　theory, 88
Concentration
　development, 99
Conservation easement, 46
Contiguity metric, 101
Continuity, 12, 99, 108
Corridor, 11–12, 14, 44
Country-living, 18, 27
CUF, 116

D
Decision tree, 66, 77–79, 121
Density
　metric, 100, 106
　population, 5, 24
　urban, 8, 10, 12–14, 29, 41–42, 99
Deterministic, 62–63, 109, 117
Development
　right, 46
　tax, 18, 23
　policy, 18, 26
Difference image, 68–69, 71, 73, 76
Diffusion, 15, 93, 111, 114
DINAMICA, 117, 120
Discontinuous development, 8, 13, 21, 26, 103
Dissection, 12
DUEM, 117

E
Earth observation, 50
Econometric panel, 66, 79
Economic
　cost, 29
　development, 38
　growth, 18, 21
Ecosystem, 6, 30

Edge development, 11
Empirical model, 62–63
Energy inefficiency, 30
Entropy, 104–106
Environmental protection, 38
Expectations, 2, 9, 18, 20–21

F
FCAUGM, 117, 120
Fractal geometry, 59, 88, 116
Fragmentation, 10, 12, 31, 88, 92, 101–102
Frame-independence, 134
Fringe development, 11

G
Geographic information system, 56
Geographically weighted regression, 74, 94
Geospatial indices, 103
Green belt, 46
Green development, 38

H
Housing investment, 18, 27
Human geography, 2

I
Image
    classification, 51, 53, 62, 77, 79–82, 125–127
    differencing, 66, 68
    index, 70
    overlay, 65–68
    ratioing, 66, 70
    regression, 66, 73
    subtraction, 68
Incentives and disincentives, 45
Independence of decision, 18, 20
Industrialisation, 18, 20–21, 28
Infill, 10–13, 22, 40, 47, 101
Infrastructure concurrency, 46
Infrastructure cost, 29
Instantaneous field of view, 53
Intensity-hue-saturation, 78
Isolated development, 11

L
Land hunger, 18, 22
Land-cover, 4
Landscape
    ecology, 87
    metrics, 87, 90
    structure, 88
Land-use, 4
Large lot, 18, 27

Leap-frog development, 12–14
Legal dispute, 18, 22
Linear branch, 10–12, 24, 25
Living cost, 18, 23
Living space, 18, 24
Low-density development, 12, 14, 29, 35, 124
LUCAS, 118

M
Mask, 69
Measurement vector, 75
Median household income, 23
Metrics, 87, 100–101, 127
Migration, 18, 19
Mitigation ordinance, 46
Mixed class, 53, 55
Mixed pixel, 53, 55, 126
Mixed use, 39–42, 99
Mixel, 53
Model
    deterministic, 62–63, 109, 117
    empirical, 62–63
Modelling, 88, 95, 107
Modifiable areal unit, 105, 130–134
Monocentric, 13–14
Multinucleated, 13

N
Neighbourhood, 42, 44
New-urbanism, 40
Nuclearity, 99
Nucleus family, 18, 26

O
Object-oriented, 80
Outlying, 10–11

P
Patch, 87
    dynamics, 87
    metric, 61, 88, 90
Pattern
    quantification scale, 125
    summarisation scale, 126
Perforation, 11
Peripheral open space, 100
Phenology, 82
Photomorphic region, 130
Physical geography, 18, 22
Physical terrain, 22
Planning policy, 18, 26
Polycentric, 13
Population growth, 18

Post-classification comparison, 66, 79–82, 127
Principal components analysis, 66, 74
Property cost, 18, 23
Property tax, 18, 23
Proximity, 99
Public
    acquisition, 44
    health, 34
    regulation, 18, 24

Q
QUEST, 117

R
Radiometric resolution, 52, 54–55, 82–83
Remote sensing, 49
Resolution, 52
Restraining policy, 45
Ribbon development, 12, 100
Road width, 18, 24

S
Scale dependency, 123
Scale effect, 132
Scatter development, 13, 100
Shannon's entropy, 104–106
Shrinkage, 11
SIMLUCIA, 117
Simulated neural network, 76
Simulation, 60, 95, 107
Single-family home, 18, 25
SLEUTH, 117, 119
Smart growth, 39–44
Social development, 38
Social health, 35
Spatial
    autocorrelation, 92–93, 97, 131
    characterisation, 130
    dependency, 131
    domain, 128
    frequency, 61
    heterogeneity, 93
    interaction, 94
    interpolation, 94
    metric, 87–91, 97, 127–128
    range, 61
    regression, 94
    resolution, 53–54, 124
    sampling, 132
    scale, 124
    statistics, 92, 96–97, 130
    structure, 88
Spatio-temporal scale, 61
Spectral pattern recognition, 80

Spectral resolution, 55
Spectral-temporal classification, 66, 72
Speculation, 18, 21
Sprawl
    causes, 17
    characterisation, 10
    consequences, 28
    definition, 8
    dimensions, 99
    metrics, 99
    pattern, 12
    process, 14
    quantification, 97
Strip development, 14, 103
Sub-pixel, 51, 53
Suburbanisation metric, 100
Sustainable development, 37
SVM-CA, 120
Switching regression, 94, 96
System, 6

T
Temperature, 31–33
Temporal
    frequency, 61, 132
    range, 61
    resolution, 52, 55, 119
    sampling, 132
Threshold, 69
Town, 6
Transition matrix, 80–81, 86
Transit-oriented development, 42
Transportation, 18, 24, 42

U
UES, 117
Urban
    analysis, 57–58, 96
    area, 3, 5
    core, 100
    cover, 4
    development, 2–3
    dynamics, 110
    ecosystem, 6
    expansion, 11
    extent, 5, 100
    footprint, 100
    fringe, 100
    geography, 1–2
    heat island, 32
    land-cover, 4
    land-use, 4
    model, 107
    modelling, 88

poverty, 28
sprawl, 7–10, 12
system analysis, 58
Urban growth, 3
analysis, 59, 85, 96
boundary, 46
causes, 17
consequences, 28
control, 44
model, 107
pattern, 10
process, 14
Urbanised area, 100

Urbanisation, 3
Urbansim, 117

W
Water quality, 34
Wealth, 30
What if?, 116
Wildlife, 30

Z
Zoning effect, 132
Zoning policy, 26